MISSING LINKS

EVOLUTIONARY CONCEPTS & TRANSITIONS THROUGH TIME

Robert A. Martin
Department of Biological Sciences
Murray State University

JONES AND BARTLETT PUBLISHERS
Sudbury, Massachusetts
BOSTON   TORONTO   LONDON   SINGAPORE

World Headquarters
Jones and Bartlett Publishers
40 Tall Pine Drive
Sudbury, MA 01776
978-443-5000
info@jbpub.com
www.jbpub.com

Jones and Bartlett Publishers Canada
2406 Nikanna Road
Mississauga, ON L5C 2W6
CANADA

Jones and Bartlett Publishers International
Barb House, Barb Mews
London W6 7PA
UK

Production Credits
Chief Executive Officer: Clayton Jones
Chief Operating Officer: Don W. Jones, Jr.
Executive V.P. & Publisher: Robert W. Holland, Jr.
V.P., Design and Production: Anne Spencer
V.P., Sales and Marketing: William Kane
V.P., Manufacturing and Inventory Control: Therese Bräuer
Executive Editor, Science: Stephen L. Weaver
Managing Editor, Science: Dean W. DeChambeau
Senior Production Editor: Louis C. Bruno, Jr.
Marketing Manager: Matthew Bennett
Marketing Associate: Matthew Payne
Text and Cover Design: Anne Spencer
Illustrations: Elizabeth Morales
Printing and Binding: Malloy
Cover Printing: Malloy

Library of Congress Cataloging-in-Publication Data
Martin, Robert A., 1944–
  Missing Links / Robert A. Martin
     p. cm.
  Includes bibliographical references and index.
  ISBN 0-7637-2196-4
  1. Vertebrates—Evolution. 2. Vertebrates, Fossil. 3. Evolution (Biology) I. Title.
  QL607.5 .M36 2003
  566—dc21   2002041049

Printed in the United States of America
07 06 05 04 03   10 9 8 7 6 5 4 3 2 1

On the cover: Archaeopteryx lithographica, the oldest known bird, is an
example of a generalized "missing link" between early theropods and birds.

*In fond memory of Stephen Jay Gould (1942–2002),
whose thoughts on natural history and evolutionary biology
inspired a generation of paleontologists
and educated a world.*

Brief Contents

Contents

Preface

WHEN DARWIN PUBLISHED his *Origin of the Species* in 1859, very little fossil material was known. Darwin himself collected many fossil specimens on his voyage around the world on *H.M.S. Beagle,* but they were primarily described by Richard Owen, an English paleontologist who rejected Darwin's mechanism of natural selection for evolutionary change. Owen was a pretty active fellow; he also named the Dinosauria. In the United States, the first fossils were gathered from a marsh at what is now Big Bone Lick State Park in Kentucky in 1739 by a French Canadian officer, Baron Charles de Longueuil. It is a strange and sometimes ironic aspect of evolutionary science that some of those influential in the development of the modern synthesis of evolutionary theory themselves did not embrace evolution or its primary mechanisms, either because they were unaware of it or simply rejected it. A few of these include the following: Carolus Linnaeus, a Swedish physician and father of animal and plant classification, who believed he was cataloguing examples of God's omnipotence; Baron Georges Cuvier, the father of comparative anatomy and paleontology, who was convinced that God initiated "catastrophes" to erase entire communities of organisms; Alfred Russell Wallace, who developed the theory of natural selection at the same time as Darwin and later rejected the idea that humans could have resulted from this process; and the aforementioned Richard Owen, England's premiere paleontologist. But the one–two punch provided by James Hutton's *Theory of the Earth* and Darwin's *Origin of Species* sent shock waves through the scientific community and began a modern exploration for fossils and their meaning that continues to this day. Hutton's work, published in 1785, was an influential one for Darwin, because it provided circumstantial evidence for an ancient Earth. Fundamentalist Christians of Darwin's time fought Darwin very strongly, because a literal reading of the Christian Bible is incompatible with both an ancient Earth and the concept of evolution. It is an amazing testimony to the power of religious influence that millions of Americans today still follow this basic credo. And one of the tenets of those who support a "creationist" model for humans and the universe is that there are no true "missing links," extinct organisms that connect modern and ancient life in an unbroken chain through long periods of time, measured in the millions and billions of years. But we have come a long way since Darwin's day. Thousands of expeditions have unearthed literally millions of fossils, today housed in many museums and academic institutions worldwide. Paleontologists have described fossils over 3 billion years old, and organisms are known continuously from that point onward. Many missing links are recognized, and it is the primary purpose of this book to provide a compendium of this information for the general reader in the context of modern scientific inquiry.

Missing Links is designed to satisfy two needs. First, as represented by the case histories, it can be read purely as a compilation of fossil histories in support of evolutionary theory. It is my hope and intention that this anthology is used in every conceivable circumstance where such information is needed, from a student classroom presentation to an intellectual religious discussion. The second usage, represented by Section I, is as a primer on evolutionary science, particularly as it applies to fossil materials. Evolution can no longer be expressed by the simple phrase "survival of the fittest." Hundreds of scientists have worked carefully for almost 150 years to establish evolution as a theory that is profound in scope and universally supported, and more scientists are likely at work today on evolutionary problems than were active in the first 125 years combined.

Evolutionary science encompasses many fields and can be extremely complicated to the novice. Consequently, I have prepared a synthesis in Section I of the major concepts, processes, and vocabulary necessary for a reasonable understanding of evolution, particularly as it applies to interpretation of the fossil record. Because the focus of this treatment is on transitions in the fossil record, there is little here on evolutionary genetics as practiced with extant organisms. For those interested in this area, I recommend the sections on that topic in the recent text *Evolutionary Analysis* by Scott Freeman and Jon C. Herron (2001, Prentice Hall). Although it is not necessary to be familiar with the contents of Section I to consult the case histories, it will be of considerable benefit. This section begins with a chapter on how science operates as a method of knowing. Most people fail to understand that science is not limited to chemistry, biology, physics, and so on, the so-called hard sciences. Everything we do as an adult is based on a series of experiments we performed as we were growing up. We could not function if this were not so. Hopefully, as this universal aspect of science becomes more widely known and appreciated, perhaps science will not seem so mysterious. Chapter 1 also documents the ancient age of the Earth and its dynamic nature. One of the greatest and most powerful discoveries of our time, the theory of continental drift, is reviewed. This is followed by a more specific examination of methods for dating rocks and fossils, called *chronometry*. Chapter 2 begins with an examination of the nature of species and then progresses to a discussion of natural selection and the variation in organisms on which this selection operates. This *microevolutionary* perspective is contrasted with patterns on a longer scale of time that we identify as *macroevolution*. Can the mechanism of selection on individuals explain all of life's patterns? As we shall see, we owe our existence to the extinction of dinosaurs by a stray extraterrestrial object, not to any inherent evolutionary program or progression toward superiority.

The publication of German systematist Willi Hennig's book, *Phylogenetic Systematics,* in 1966 led to a major revolution in the way classification is accomplished. Today, with the help of powerful computer programs such as MacClade and PAUP, we methodically examine and analyze character variation in modern and fossil organisms to identify relations among them. The underlying philosophy of phylogenetic systematics, or *cladistics* as it is sometimes known, is that classification should represent phylogeny or the revealed genealogy of organic life. The basic principles of this approach are outlined in Chapter 3.

Chapter 3 also considers the myriad influences that lead to the creation of a fossil locality and the way scientists learn about past environments. What forces affect the burial, preservation, and distribution of fossils? In what ways can fossils be used to interpret past climates? In addition to these applications, fossils also can lead to some profound ideas about how ecological communities change through time. One of the more controversial of these ideas, suggested by University of Chicago paleontologist Leigh Van Valen, is that species become extinct at a constant rate through time, never quite able to keep up with changes in the environment around them. The mechanism of constant extinction is known as the "Red Queen Hypothesis," named after the famous character in Lewis Carroll's *Through the Looking Glass.*

The case histories begin, in Chapter 4, with a scenario for the appearance of life on Earth (with contributions from Mars, if that turns out to be the case). We see how Stanley Miller and others have generated all the building blocks of complex cellular molecules, such as DNA and ATP, in the laboratory, and how these results are consistent with what we know of the early atmosphere and origin of life as indicated in the fossil record.

Chapter 5 tells the magnificent story of whale evolution that has unfolded only in the last 10 years, complete with aquatic intermediates sporting tiny hooves on their feet. In Chapter 6 we look at the vast Cenozoic panorama of horse evolution, showing the eventual progression of changes leading from tiny forest dwelling ancestors to the large and speedy descendants of today. Thanks largely to the modern work of Bruce MacFadden and Richard Hulbert, we now can see that horses were a diverse lot, with many experiments in both size and morphology that did not survive.

Archaeopteryx needs little introduction, but again, as in whales, there have been many new discoveries linking early birds to their modern relatives. Chapter 7 examines these new finds in the context of a raging debate about bird ancestry. Some paleontologists, such as Robert Bakker, Louis Chiappe, and Kevin Padian, are convinced that birds arose from dinosaurs and, in fact, are dinosaurs, whereas another camp, led by Alan Feduccia and Larry Martin, believe this perspective will turn out to be a great embarrassment. The latter two are convinced that birds originated from an earlier reptilian ancestor that may have given rise to both dinosaurs and birds. Whatever the outcome, new fossils have cemented the links between reptiles and birds. Lots of early "toothy" intermediates are now known. Chapter 8 reviews the fossil evidence for the transition from reptiles to mammals. It is an interesting segue, involving not only modifications in thermoregulation but also in hearing and food acquisition and processing. Were the last mammal-like reptiles endothermic? Did they possess fur? How does one define and recognize a mammal? Examination of the fossil record will show how this question is answered with the fossil material available.

I and others have chosen to work with the fossil history of the innocuous but ubiquitous rodents, in the hope that the dense fossil record of these animals will reveal the secrets of evolution, if for no other reason than so many fossils are available for study. Chapter 9 focuses especially on microevolution, that is, changes within species lineages over long periods of time. In rodents we have good fossil evidence for considerable morphological and size change at different chronological scales. Here we can get a close-up look at how links operate, and we see that it is often a blurry mess of related populations originating in an almost helter-skelter fashion, each with its own set of characteristics and evolutionary trajectories.

In Chapter 10 we examine one of the great changes in the history of vertebrate animals, the transition from water to land, or *terrestrialization*. Traditional ideas on the origin of limbs in land animals, such as possessed by the modern amphibians, proposed that hands and feet evolved to support the body on land. Not so, says Jennifer Clack of Cambridge University. Her work indicates that tetrapod (four-footed animal) limbs appeared first in fully aquatic animals. If limbs are not present for support on land, then what was their original purpose? This and other topics relating to the movement of our ancestors onto land are considered in Chapter 10.

Our own heritage is taken up in Chapter 11. As with the whales and birds, new hominids are being reported constantly. As I began outlining this chapter a few years ago, *Australopithecus bahrelhghazi* was described from three to four million year old deposits in Chad. The previous year saw the description of *Ardipithecus ramidus,* perhaps the sister group of the lineages leading to modern humans. I had to rewrite Chapter 11 two weeks before the manuscript was due in the publisher's office because of the announcement of the amazing *Sahelanthropus* from central Chad. It is an exciting time in paleoanthropology, and as one might expect, there are probably as many theories of human relationships as there are investigators. An old joke suggested there are likely as many anthropologists studying hominids as there are fossils, a past testament to a lousy fossil record, but that is not true any more. There are now hundreds of specimens of *Australopithecus afarensis* alone.

The final chapter is one that may seem unusual in a book on missing links, but hopefully by the time the reader has made it this far its purpose will be obvious. In this chapter I consider the "smoking gun" of evolution: evolution in action during the lifetime of the reader as an observer. Missing links on a generational scale, complete with the sudden appearance of new forms, anatomical change, and the potential for new species.

R.A.M.

Acknowledgments

I must first acknowledge evolutionary scientists worldwide, whose diligent research provided the source of most of the information in this book. I could not cite even a small part of the tremendous written database supporting Darwin's grand theory of evolution, and I apologize to the hundreds of scientists, past and present, whose names do not appear here.

Many of my colleagues contributed in one way or another to the writing of this treatment. Jim Honey and Pablo Peláez-Campomanes taught me everything I know about field paleontology, from identifying likely fossiliferous sites, to stratigraphy, to field procedures to reclaim and preserve the fossils. They are great friends and field companions, and I hope they understand the extent of my appreciation. Jim Farlow provided a detailed review of an early manuscript, and I hope he is not too annoyed that I decided not to publish with his university press. Ted Daeschler, Robert Carroll, Jim Hopson, Richard Hulbert, Jr., Hans Thewissen, Larry Martin, Jennifer Clack, Christopher Sloan, and Kevin Padian either reviewed chapters or provided useful information and opinions. Thanks are also due Ian Tattersall for granting permission to use some of the many outstanding illustrations in his book, The *Fossil Trail*. I tried out early versions of *Missing Links* in my undergraduate Evolution classes at Murray State University, and I appreciate the help provided by Tom Timmons and Barbara Covington in making the various accommodations necessary for this process. Crystal Rose and Meredith Duguid, students at Murray State University, helped track down information and prepare the illustrations.

Judy Hauck, a former editor at Jones & Bartlett, was the first individual in the publishing business to get excited about *Missing Links*, and I want to thank her and the current science editor, Stephen Weaver, for their positive encouragement. My appreciation is also extended to the editorial and production staff at Jones and Bartlett, including Dean DeChambeau, Rebecca Seastrong, Anne Spencer, and Lou Bruno. My family, Linda, Alice, Rachel, and Jeremy, have been my biggest fans, and in actuality this book should be dedicated to them as well as to Steve Gould.

Introduction: What Is a Missing Link?

In colloquial language a "missing link" is generally considered to be a connecting species between higher taxonomic categories, such as between a fish and an amphibian or between an ape and a human. Most people tend to think of a missing link as a single kind of animal, at the exact midpoint between two quite different animal groups. It is sometimes thought of as a special kind of organism that represents a "quantum leap" in evolution. In fact, links between major organismal groups were no different from any evolutionary connection in life's history, except by virtue of what came after them.

Every living organism is a member, at the same time, of a breeding group, a population, and a species. Humans are members of the species *sapiens* in the genus *Homo*. Conventionally this is expressed as the species *Homo sapiens*. The genus is a level that includes related species, all of which presumably evolved from a common ancestor. A species includes only individual organisms, arranged in populations of breeding groups. Genera (the plural of genus) can be grouped into families, families into orders, orders into classes, classes into phyla, and phyla into kingdoms. These represent higher levels, or scales, in a standard organismal classification. Humans appear to have evolved from apes; this represents a link at the family scale—between the Pongidae and Hominidae (if we follow a classification that treats these as separate families; some do not). Both families are in the mammalian order Primates. The link between dinosaurs and modern birds represents a considerably more distant connection. It seems there should be something different about links at these different scales, but it is just an illusion.

Individual living humans have unique family histories, but they eventually blend with other humans as we go back far enough in time. Every human alive today is likely descended from a single human ancestor at some point in the past. So there are links even within the single human species. The history of parental and offspring descent that represents the genealogical tapestry of

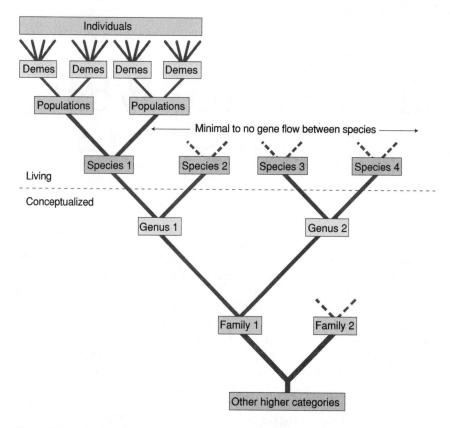

Figure I.1 A genetic/hierarchical organization that represents living (above the dashed line) and conceptualized associations. The conceptualized associations, of course, if interpreted correctly and including information from the fossil record, can provide the basis for a classification and a history of genetic connections among related groups.

humans is known by biologists as a *lineage*. Many lineages are known in the fossil record, and some are millions of years old. It is now understood that populations of individuals making up a lineage can change, or evolve, through time. We call this kind of change *phyletic change,* or *phyletic evolution.* So, how does this all bear on the concept of a missing link?

First, it is important to realize that all taxonomic units above the level of the species are artifacts of our need to make order out of the universe. A "genus" is not a real biological unit, nor is a "family." These terms simply ally sets of related species in an ever-enlarging family tree. The names become more inclusive as we progress farther back in time and include more common ancestors and their descendants, until we have created a complete family tree and named, for our own convenience, a variety of descriptive and helpful higher taxonomic categories (**Fig. I.1**). *Thus links, missing or otherwise, can only be individuals.* There is nothing else in nature. We do not look for the link between ancient fishes and modern amphibians (frogs, salamanders, etc.) to be qualitatively any different than the link between apes and humans, despite the fact that the links in which we are interested are at different taxonomic scales. Expressed another way, links include only two categories: the individuals in a lineage (= a species) and the individuals that connect different lineages.

Furthermore, to the extent that lineages change, a missing link is not just a single organism, but all the individuals within the species that change. Thus, we often identify *Archaeopteryx lithographica* as an example of a missing link between early theropods and birds, but there were likely many hundreds or even thousands of individuals with slightly less and slightly more bird-like features in the link from dinosaurs to birds. Some existed within one lineage and others in different species that make up the full connection between dinosaurs and birds.

We might believe that there must at least be a *quantitative* difference between links at different taxonomic scales, maybe more species at the higher scales. Perhaps there are more species linking dinosaurs with birds, or fishes with amphibians, than apes with modern humans. But there is no evidence for this of which I am aware, nor any reason why it should be so. Each transition includes individuals within species and must be considered on its own merits, taking place under its own unique environmental conditions. Actually, the number of links one identifies can depend on the way the particular question is phrased. For example, the system defined by the question "What are the links between fishes and modern humans?" obviously includes more links than one defined between fishes and amphibians, for the simple reason that in the former case we are asking a question about two groups that did not directly evolve into one another, but included a series of groups in between.

In the 1950s a biologist named Richard Goldschmidt suggested that the origin of evolutionary novelties leading from one major kind of organism to another could have appeared instantaneously, in a single generation, as a result of mutations in developmental instructions. This idea came to be known as the "hopeful monster" scenario. It has appeared in various guises in the literature; another expression, "quantum evolution" means basically the same thing. The problem with this model is that almost all major changes of this sort with which we are familiar are lethal; offspring expressing major deviations from their parents generally do not survive because they are less fit for the environment into which they are born. In theory, hopeful monsters do not make sense. The probability that a radically new anatomical design would fit in better to a set of environmental conditions than the well-adapted parental one seems to be very small, though given the immense time since the earliest

Figure I.2 *Pikaia gracilens,* an early chordate from the Cambrian Burgess Shale of British Columbia. (From J.H. Ostrom, 1992. A History of Vertebrate Successes. In J.W. Schopf, ed. *Major Events in the History of Life.* Sudbury, MA: Jones and Bartlett. Photo Courtesy of S. Conway Morris.)

appearance of life, it probably has happened occasionally. Most species that we recognize today as important connections in their particular evolutionary chains were, at the time they existed, generally no more remarkable than their parents and the species that came before and after them. We say that *Archaeopteryx lithographica* is a good representative link between early theropods and birds only because we have the entire later history of avian evolution. Without it, *Archaeopteryx* would be just another small bipedal reptile with feathers. *Pikaia,* a likely 500 million year old ancestor of all later chordates, including humans, looked like an unobtrusive worm-like creature in its near-shore marine environment (**Fig. I.2**). There were many more interesting and bizarre (from our perspective) creatures that shared its world. But most of them became extinct, and only *Pikaia's* descendants survived and eventually stood on the Moon.

1

Sources of Knowledge and Earth History

"A belief is simply a hypothesis in need of an experiment."

There Are No False Clues: Science as a Way of Knowing

There is no such thing as a bad idea. Some ideas are certainly more interesting than others, or more easily testable, but any independent idea is a worthy thought. In America today, as well as in the rest of the world, there is a raging controversy over two ideas that have been around a long time. The first is evolution, and the other is creationism. Evolutionary theory holds that all organisms, including humans, evolved from a common ancestor billions of years ago. Creationism, on the other hand, proposes that all life on Earth was created by a supernatural being called God. Christian fundamentalist creationists further assert that the Christian Bible should be read literally, and, therefore, the age of the Earth is very young, perhaps less than 10,000 years old, and that the current diversity of life on Earth can be traced directly to a time after a worldwide biblical flood. How do we determine which idea is correct? For fairly obvious reasons, both are not likely to be valid. The Earth is either very young or it is not. But how can we *know* which is correct? The resolution of this controversy requires the same techniques that we would use to gain basic knowledge about any subject. In the following sections I endeavor to show that the approach we use to know anything with certainty is always the same and that this technique is used by everyone from the moment of birth until death. We call it the *scientific method*, but it is practiced by everyone. Science is a way of knowing and has nothing to do with a specific discipline of study.

The Method of Science

Millions of people are afraid of science and mathematics and so choose not to learn about it. To many, science is for "nerds," and when they think of science and scientists, they get the stereotyped vision of a bespectacled man with frizzy hair and a beard pouring a foaming substance into a beaker, giggling madly. The chief scientist/financier at Jurassic Park, ably played by Sir Richard Attenborough in Steven Spielberg's rendition of a Michael Crichton novel by the same name, was a bit more kindly and high-tech, but nonetheless came across as a misguided senseless fool (and in the book the dinosaurs ate him). In all fairness, the dinosaur paleontologists, played by Sam Neill and Laura Dern, came off a bit better, but never in the mass media (film and television) does the public ever get the message that everyday people are scientists and that the only difference between the characters played by Neill and Dern and them is the careers they choose. Some scientists choose to become chemists, biologists, and paleontologists. Others become historians, accountants, and mechanics. We are all of us scientists, because science is not what we are, but what we do. We do not always do science, and when we do not bad or unpredictable things often happen, or we believe things that cannot be true and in so doing develop a life-style destined to alienate us from a large segment of society. So what then is science if it is not biology and chemistry and physics? And what does this all have to do with missing links? Science is a *method of knowing* about our world that allows everyday scientists to survive and more highly trained scientists to discover secrets of the history of life, like missing links.

Science begins with *observations*. Like a teenager learning to drive. Observation: An automobile is seen to turn right and left. Observation: A steering wheel can be turned clockwise (right) and counterclockwise (left). These observations then lead to an idea: If the steering wheel is turned counterclockwise, the car may turn to the left; clockwise, to the right. In science, this idea is called a *hypothesis*. The idea must now be tested to see if it is true. Scientific tests are called *experiments*. Once in the car, the teenager turns the steering wheel to the left and, voila, the car turns to the left. This has been an experiment, and if the experiment, when repeated, turns out the same every time, then the hypothesis has been confirmed and the teenager (experimenter) can conclude that under all similar circumstances the hypothesis will be true. As far as I know, this mode of turning a car exists everywhere on the planet. Consequently, the teenager can take the hypothesis one step further and say that it is so well documented as to be a general *theory*, like atomic theory or the laws of thermodynamics. Even in a strange place, such as a new country, steering wheel and car turning will be coordinated. This is science in action. If we did not do science every day, we could not function. Science is required to open doors and cans, walk down stairs, button a shirt, flush a toilet, and turn on a light. However, sometimes we either do not do science or we do it badly. That is when we end up without knowledge and often frustrated, misguided, or mistaken. Let us look at an example.

Linda sees (observation) her husband John with his arm around Sally. She assumes (hypothesis) John is having an affair with Sally. She confronts John and accuses him. He denies the accusation and instead claims he was helping Sally because she had a sprained ankle. What is the truth? We do not know, because a critical part of the process is missing, the experiment. The hypothesis was not tested. A conclusion was reached without data (information). Now, Linda may think she knows the truth, and that may lead her to act in a certain way toward John, but if she does it will not be based on her knowing anything for certain, it will be based on a *belief*. This example is provided to distinguish between knowledge obtained by the scientific method and "knowledge" obtained by a belief system. One does not gain knowledge from a belief; a belief is simply a hypothesis in need of an experiment. Let us examine this in a bit more detail with the evolution/creationist ideas.

The creationist model is a valid hypothesis for the origin and diversity of life on Earth, as is the evolution model. To provide support for one model over the other, we must set up a test, or an experiment, in which we would make observations that either support or deny one model versus the other. Fortunately, this is not too difficult to do, because each model makes certain predictions about the world and the organisms in it. Some of them would be as follows:

Creationism	Evolution
1. The Earth is young, probably around 10,000 years old.	1. The Earth is billions of years old.
2. The universe is young.	2. The universe is billions of years old, and the age of the Earth is consistent with the age of the universe.
3. There will be no special order to fossils; one can expect mixtures of animals and sedimentary rocks from the flood. Humans coexisted with dinosaurs.	3. The fossil record will show a distinct order. There will be no contradictory evidence, such as humans and dinosaurs together.
4. Evolutionary links between ancient and modern organisms cannot exist.	4. Evolutionary links between ancient and modern organisms must exist.

continued

Creationism	Evolution
5. Humans will not share much in common biologically with other organisms.	5. Humans will share many basic biological features with other organisms, especially with their closest living relatives, the Primates, and other vertebrates.

A considerable body of evidence supports the evolutionary model that comes from a diverse set of disciplines: geology, paleontology, ecology, developmental biology (embryology), biochemistry, genetics, and physics (cosmology). Repeated testing of the age of the Earth and other objects in space (the Moon, meteorites, etc.) indicates the Earth and the universe are indeed billions of years old. A sedimentary record of life on Earth goes back billions of years, and there is a comprehensible progression of organisms with no contradictory evidence (e.g., no dinosaurs with humans). Humans and chimpanzees share 98% of their DNA; they share less with less closely related animals. All animals have many genetic and structural similarities (**Fig.1.1**). There is no reasonable explanation for this other than that they share a common ancestry. If there had been a single worldwide flood and if the world was young, then one would expect no ordering to fossils found in sedimentary layers. There would not be distinct layers with only fishes, followed by layers with fishes plus amphibians, and so on. Everything would be a complete mix. Single floods cannot produce, as they do in the Grand Canyon, ancient sediments more than a mile thick that range over a billion years with steadily more complex organisms in the rock units from the bottom (oldest) to the top (youngest). This book is concerned primarily with the "first-order" evidence for evolution, the fossil record. After all, it is only with the fossil record that we can possibly see major patterns of change over long periods of time. The creationist/evolution models are clearly distinct here. If the creationist model is correct, there can be no ancestor–descendant relationships, no evolutionary progressions of any kind. If the evolution model is correct and with a dense enough fossil record, intermediates at all structural and chronological scales should be found.

Paradoxically, knowledge obtained by the scientific method is never certain. There is an interesting bumper sticker that says "Question Authority." It is meant as a political statement; never trust people in power, but it can be aptly transferred to any claims of knowledge. Our senses, even with the extensions of technology, remain extremely limited. By repeated experimentation it is conceivable that we have come across some of the fundamental

Stage 1

Stage 2

Stage 3

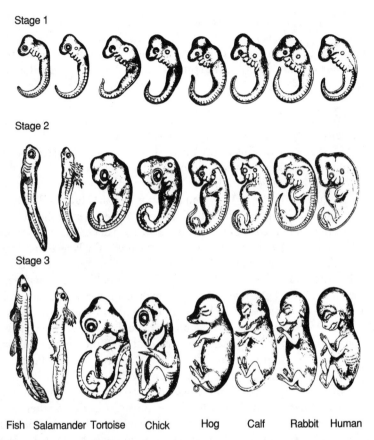

| Fish | Salamander | Tortoise | Chick | Hog | Calf | Rabbit | Human |

Figure 1.1 Similarities among vertebrate embyros are considered as evidence of close phylogenetic relationship. Three stages in embryonic development of various vertebrates are illustrated with humans at the far right. Note the many similarites during the first two stages. (From J.G. Romanes, 1910. *Darwin, and After Darwin.* Chicago: Open Court Publishing.)

universal truths, but we must constantly be wary and always be willing to try new experiments, even on world class theories. As we shall see, the best scientific minds of the twentieth century rejected Alfred Wegener's hypothesis of continental drift as recently as the 1950s, but he was right. We must always couch our current scientific knowledge as "currently unfalsified." Tomorrow may be different. But it is this very feature of science that ensures the health and honesty of a society and determines that we will continue to progress intellectually as a species and know something profound about the universe, perhaps even how it works and where it came from. Where would we be today if religious dogma of the fifteenth century had prevailed and everyone had accepted that the stars, the sun, and the planets revolved around the Earth? It seems like a ridiculous notion, but it is not so far from home. If modern Christian fundamentalists had their way, evolution would be banned from textbooks and only the Biblical theory of Genesis and a young Earth would be taught to American children. But Genesis is a parable; we have a pretty fair idea of how life appeared on Earth, and when, and what happened subsequently.

Science and Cosmology: The Age of the Universe

People who study the origins of the universe, often coupling astronomy and nuclear physics, are known as *cosmologists*. They tell us that the universe is about 12 to 15 billion years old. This number is deduced, in part, from observations made by the great American astronomer Carl Hubble. The Hubble telescope currently in orbit around the Earth was named in his honor. Hubble's critical contribution to astronomy and our knowledge of the universe was based on the observation that almost every star in the night sky is receding from us at a rapid pace and that the farther the star is from us, the faster it is receding. Hubble determined what has since been called the *Hubble Constant,* that for every one million light years distant from an observer on Earth, an object increases its velocity by about 15 km/sec. One might wonder how this is possible if the Earth is not the center of the universe, but it is very simple if one can imagine the universe expanding from a central point after an unimaginable explosion commonly referred to as "The Big Bang." (Arno Penzias and Robert Wilson, formerly of Bell Labs, won the Nobel Prize in Physics for their discovery of the background radiation from this blast, which had been predicted by Albert Einstein many years previously.) Imagine further that the universe in these initial moments can be modeled by a deflated balloon. If we draw dots on the balloon's surface to represent many of the stars in the universe and then follow the pathways of those dots as the balloon expands, we discover that space develops between all the dots. That is, to an observer standing on any star (or any planet revolving around any star), it will appear as if all other stars are moving away in all directions. (Today some cosmologists believe the universe is actually flat; a bizarre notion, but the balloon analogy would still apply to any particular point.) We gain support for this model by characteristics of light produced by stellar objects. The light produced by an object is actually a spectrum of many colors, ranging from a deep indigo blue at one end to red at the other. When an object is moving away from an observer, the light waves produced are spread out and are seen as shifted toward the red end of the spectrum. Conversely, light produced by an object approaching an observer is shifted toward the blue end of the spectrum (this is a form of the *Doppler effect,* originally described for sound). Hubble knew that the stars were moving away from us and therefore the universe was expanding, because everything in the night sky was red shifted (**Fig.**

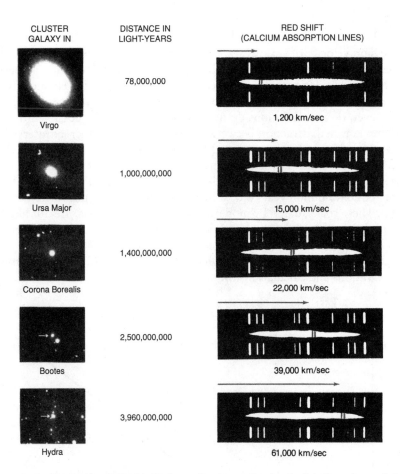

CLUSTER GALAXY IN	DISTANCE IN LIGHT-YEARS	RED SHIFT (CALCIUM ABSORPTION LINES)
Virgo	78,000,000	1,200 km/sec
Ursa Major	1,000,000,000	15,000 km/sec
Corona Borealis	1,400,000,000	22,000 km/sec
Bootes	2,500,000,000	39,000 km/sec
Hydra	3,960,000,000	61,000 km/sec

Figure 1.2 The cosmological redshift refers to the observation that the light from almost all distant objects in the night sky is shifted towards the red end of the spectrum. As noted in the mid 1800s by the Austrian physicist Christian Johann Doppler this has to do with the stretching of waves (either sound or light) as objects recede from an observer. As objects producing sound or light move toward an observer, the waves they produce become compacted, and, with regard to light, we see this as a blueshift instead.

The American astronomer Carl Hubble was the first to determine that the extent of the redshift of objects is related to their velocity, and through studies of the light produced from distant stars and galaxies, he discovered further that those objects most distant are also moving away from us at the fastest speeds. Because there is a constant rate of speed increase per unit distance (10–15 km/second per million light years), astronomers label this as the Hubble constant. Uncertainty in the value of the constant is the result of conflicting results from different analytical methods. Naturally, if virtually everything we see is redshifted, then this provides important evidence that the universe is expanding. In this illustration, representing the calcium absorption lines in light spectra from various distant galaxies, we can see that Virgo is much closer to us than is Hydra. (Adapted and modified from R. Jastrow, and M.H. Thompson, 1972. *Astronomy: Fundamentals and Frontiers.* New York: Wiley.)

1.2). It was from the extent of the color shift that he was able to calculate the recession velocity of various stars and thereby determine that the oldest farthest objects in the universe were racing away from us at enormous speeds.

Simple calculations based on these expansion rates, which I do not go into here, indicate that the universe should be about 15 billion years old (**Fig. 1.3**). However, these calculations do not take into account the possible influence of gravitational attraction between stars, planets, and other extraterrestrial matter, and after that is accounted for the universe is estimated to be about 12 billion years old. Is this a reasonable number? Might it be much younger? Well, it is not likely much younger. The oldest rocks on the Earth and the Moon are about 4.5 billion years old, so we know the universe must be older than that. These rocks, dated by modern chronometric methods, are not likely to be off by much and therefore provide, in themselves, circumstantial support for the 12 billion year figure for the universe. If we had determined the Earth was 20 billion years old, we would have had a serious problem with Hubble's observations and calculations. But they seem to stand up. The main point of

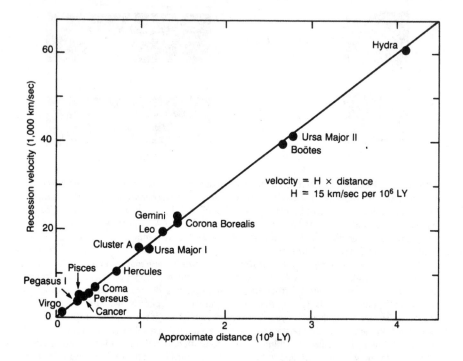

Figure 1.3 Relationship of recession velocity (speed) of galaxies as a function of distance from an observer on Earth. Assuming a Hubble constant of 15 km/sec, then a galaxy at 1 billion light years from Earth has a recession velocity of 15,000 km/sec.

all this is to show how science leads, sometimes by accidental discovery and certainly by careful observations and measurements, to testable ideas with tremendous explanatory power.

An Ancient and Dynamic Earth

Hutton and Wegener: The Earth's "Heat Engine" and Continental Drift

Darwin recognized that for evolution to have taken place, for all of life's diversity to have changed from earlier simpler forms, the Earth had to be more than the paltry few thousand years suggested by a literal interpretation of the Bible. James Hutton provided the first evidence for "deep time" through his revolutionary interpretation of geological structures and the forces that created them. Hutton and his contemporary, Charles Lyell, gave us the principle of "uniformitarianism," which simply states that the geochemical processes active now on the Earth were also active in the past. Although it is sometimes hard to imagine, the same wind and rain, freezing and thawing, and desiccation forces that act today to cause minor erosion are also the forces that, over millions of years, erode away whole mountains. Hutton was a geology professor at Edinburgh University in Scotland, and he was impressed by how little Hadrian's Wall, a rock edifice extending across much of Scotland, had changed since it was built by the Romans in 200 AD. He correctly reasoned that if it took so long to weather away a mere wall, it must take thousands or millions of times as many years to erode away entire mountain chains. He was also excited by the discovery of vertical sedimentary rock layers, many meters thick, lying under younger horizontal beds. What forces, he asked, could cause such thick rock units to warp 90 degrees? How long must such a process take? Hutton was convinced of a deep "heat engine" in the Earth that drove processes such as this. He was correct, but he could not at that time imagine the massive effects this engine could have on the Earth's topography. This was for a later generation, particularly the intrepid German Arctic explorer Alfred Wegener.

Wegener disappeared on a scientific mission to Greenland in 1930, apparently lost to an icy crevasse. He was interested in weather patterns and the nature of polar ice, and he made a number of Arctic expeditions under what we would consider today to be extraordinarily primitive conditions. As a global thinker, he was struck by the similarity of the coastal geometry of various continents, particularly Africa and South America, and he suggested that these and other continents had been joined millions of years ago in a supercontinent he called Pangaea and had later shifted apart, much as the polar ice pack does today (**Fig. 1.4**). This radical idea was met with great reluctance

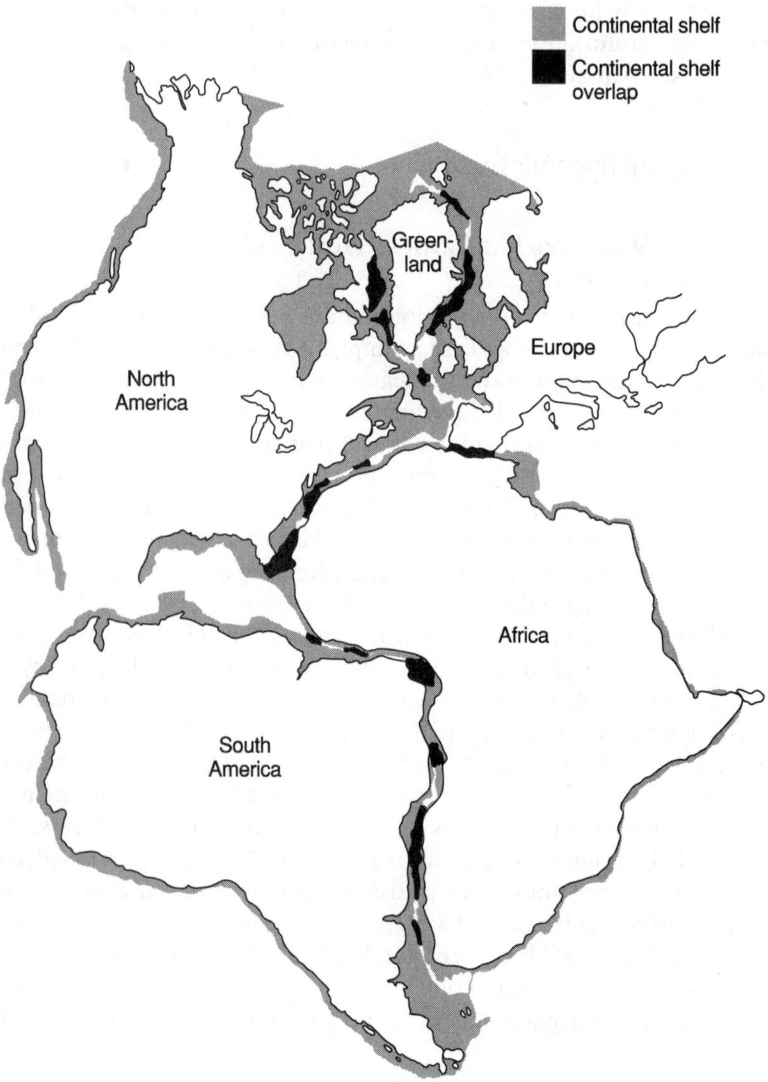

Figure 1.4 A hypothetical fit between continents according to the model of continental drift first proposed by Alfred Wegener. (Adapted from D.L. Eicher, and A.L. McAlester, 1980. *History of the Earth.* Upper Saddle River, NJ: Prentice Hall.)

by the scientific community for two reasons. First, Wegener was at a loss to provide a mechanism for such colossal movement. What could possibly cause whole continents to "move through" the Earth's crust? Second, biologists of the day were hesitant to accept the shifting continent theory as a possible explanation for the distribution of animal and plant groups. They clung tenaciously to the idea that there had been overland dispersal followed by later extinction of ancestral and geographically intermediate populations. Many of the most influential evolutionary biologists of the twentieth century attended a conference held at the American Museum of Natural History in the 1950s and virtually unanimously concluded that continental drift did not happen. But they were all wrong, as a graduate student later proved.

Paleomagnetism and Shifting Seafloors: The Birth of Plate Tectonics
Before we visit Frederick Vine at Cambridge, it is necessary to take a slight detour to examine the principle of *paleomagnetism*. Back in the 1950s, two California scientists, Alan Cox of Stanford University and Brent Darymple of the U.S. Geological Survey, made the amazing discovery that the Earth's magnetic poles had reversed position not just once but many times in the past (**Fig. 1.5**) (major periods of reversal and normality are called *chrons*). Minute pieces of magnetized metals, such as iron, orient themselves in rocks depending on the prevailing global magnetic environment. This positioning is preserved in ancient rocks and can be read today by a device called a "magnetometer." During major periods of reversal, a compass would have pointed south instead of north. Darymple and Cox confirmed their findings with studies of remnant magnetism from rocks of all ages everywhere on the planet. This fortunate discovery set the scene for the confirmation of one of the greatest scientific revolutions of all time.

Primarily as a result of a worldwide international effort to map the seafloor in the 1960s, it became common knowledge that a great ridge of mountains lay beneath the Atlantic Ocean, running in an essentially north-south direction and intersecting other ridges at various points. Many of these submerged mountains are higher than Mount Everest in the Himalayas. Seen from space, the mountain range looks like the raised seam on a baseball. The mapping program took cores of rocks from the sea bottom. Because it was assumed the most ancient rocks were to be found on the seafloor, scientists were hopeful that they would eventually find a complete record of Earth's history from these cores. Much to their surprise, the seafloor was comprised of relatively young rocks; billions of years of Earth's history were missing. But how could that be? Harry Hess, a professor of Geology at Princeton University, hypothesized that the seafloor was in fact a dynamic structure, with new floor being added near the mid-Atlantic Ridge and being subtracted at the edges. But he called his own idea mere "geopoetry."

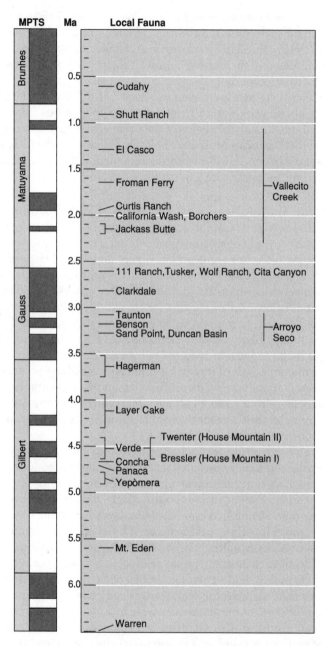

Figure 1.5 A magnetic polarity time scale (MPTS) reflects the flip-flop of the poles (compass direction) in the past. Local faunas are fossil mammal assemblages in the western United States and northern Mexico. Black - normal compass direction; white = reversed; Ma = million years ago. (Adapted from L.B. Albright, III, 1999. Biostratigraphy and vertebrate paleontology of the San Timoteo Badlands, southern California. *University of California Publications, Geological Sciences.* 144: 10.)

Frederick Vine was a graduate student at Cambridge University when he was asked to examine some paleomagnetic data that his mentor, Drummond Matthews, had brought back from the floor of the Indian Ocean. After examining these data and paleomagnetic data from seafloors elsewhere, the student and his mentor determined that the paleomagnetic signals from either side of any given seafloor ridge were mirror images of each other. Other geologists had showed previously that the youngest rocks on the seafloor were always those closest to the ridge and that the rocks got progressively older the farther out on either side one progressed from the ridge. So Vine and Matthews had finally cracked an ancient code: The seafloor was constantly being created at these underwater ridges, coming up through huge cracks in the seafloor as molten rock, or magma, and forcing the seafloor to move symmetrically away from the ridge. As the magma cooled, the rocks became magnetized according to the prevailing polar magnetic orientation, resulting in the symmetrical pattern of magnetic reversals on either side of the ridge. But where did the seafloor go at its edges? For that matter, where were the edges? Following Vine and Matthews' work, many other investigators quickly determined that the crust of the Earth was not a cooled and uniform surface riding on its molten core but rather was composed of a series of pieces, called *plates*, some of which abut each other and some of which sit slightly astride each other. As new seafloor is added at one end of a plate, the other end either pushes against another plate or is occasionally forced beneath another plate, where it melts and becomes part of the Earth's core again. This process of one plate sliding under another is called *subduction* (**Fig. 1.6**). As scientists determined the extent of the plate boundaries on the planet, they began to realize that about 90% of all volcanic activity takes place at plate boundaries, which makes sense, because this is the most accessible place for magma to come to the surface. Japan is wracked by earthquakes because the country sits astride the western boundary of the Pacific plate. Every time this great

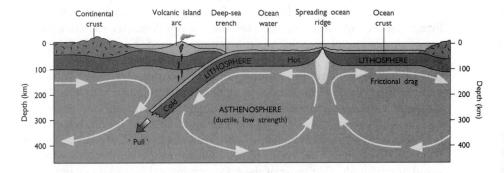

Figure 1.6 The science of plate tectonics describes the dynamics of pieces of the Earth's crust, known as "plates," as they may slide beneath other plates at a subduction zone (as in the deep-sea trench illustrated here), or they may form at mid-oceanic ridges.

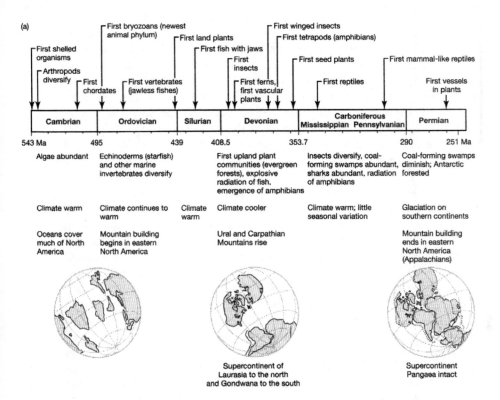

Figure 1.7 Major events in Earth history and positions of the continents through time. (*Evolutionary Analysis, 2e,* by Freeman/Herron, copyright 2001. Reprinted by permission of Pearson Education, Inc., Upper Saddle River, NJ.)

plate shifts on its journey to the core below, a volcano erupts or an earthquake occurs. A similar problem exists for the people of southern California living near the San Andreas fault. The fault line represents the eastern boundary of the Pacific plate, which is moving northward. Every time the plate moves, in distinct pulses, an earthquake occurs there as well. Los Angeles is part of the Pacific plate, and at some point, millions of years in the future, it will reside next to San Francisco. Imagine the planning that will have to go on regarding zoning, tax restructuring, and athletic contests.

The field of science spawned by the Vine and Matthews findings that deals with the formation and dynamics of plate movement is called *plate tectonics* (**Fig. 1.7**). This science has tremendous explanatory power. We now understand, for example, that most of the great continental mountain chains were formed through crustal warping associated with plate movements. Not all plates

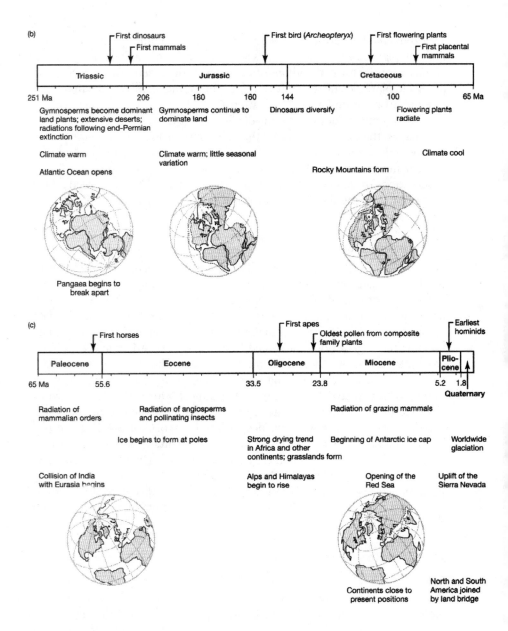

(b)

First dinosaurs
First mammals
First bird (*Archeopteryx*)
First flowering plants
First placental mammals

| Triassic | Jurassic | Cretaceous |

251 Ma   206   180   160   144   100   65 Ma

Gymnosperms become dominant land plants; extensive deserts; radiations following end-Permian extinction

Gymnosperms continue to dominate land

Dinosaurs diversify

Flowering plants radiate

Climate warm

Climate warm; little seasonal variation

Climate cool

Atlantic Ocean opens

Rocky Mountains form

Pangaea begins to break apart

(c)

First horses
First apes
Oldest pollen from composite family plants
Earliest hominids

| Paleocene | Eocene | Oligocene | Miocene | Plio-cene |

65 Ma   55.6   33.5   23.8   5.2   1.8
                                                                          Quaternary

Radiation of mammalian orders

Radiation of angiosperms and pollinating insects

Radiation of grazing mammals

Ice begins to form at poles

Strong drying trend in Africa and other continents; grasslands form

Beginning of Antarctic ice cap

Worldwide glaciation

Collision of India with Eurasia begins

Alps and Himalayas begin to rise

Opening of the Red Sea

Uplift of the Sierra Nevada

Continents close to present positions

North and South America joined by land bridge

are subducted beneath others; some are driven into other plates with essentially no place to go. Something has to give, and it is often the crust itself. The Himalayas were formed millions of years ago as the Indian plate smashed into the Asian plate. Scientists wondered for years at the fossilized remains of marine organisms found at high elevation in the Himalayas. Although the scientific community does not consider it likely that a sea once covered the Himalayas, the forces that caused the uplift of these mountains was not completely understood until the theory of plate tectonics came of age. And now, wonderfully, we have both a confirmation and an explanation for Hutton's observations of massively folded sedimentary rocks. There is indeed a great heat engine driving the Earth's surface structure.

Time Enough Indeed

In 1650, the Archbishop James Ussher of Armagh, Ireland declared that the Earth was created in 4004 BC. Approximately at the same time, another cleric by the name of James Lightfoot, then Vice-Chancellor of Cambridge University, determined that the creation occurred exactly at 9 a.m. on October 23. Both calculations had been made from various inferences in the Bible, particularly the number of genealogies assumed from the time of Adam and Eve. These attempts may seem silly and primitive to us now, but for the time it was a fairly ingenious method for figuring out one of our most basic questions: "How old is the Earth?" And it would be almost 300 years before we would have scientific methods that accurately determined the age of rocks and, thereby, fossils. Before we delve deeper into the world of fossils it is probably worthwhile to review some of the principles and methods of dating these objects and the rocks in which they are found. Two general approaches are used to date rocks and fossils. One is considered a means to ascertain the approximate true age of the rock unit, and is therefore called "absolute dating." The other approach, where fossils are dated indirectly, is known as "relative dating."

Dating in Real Time

Radioisotopes

Reasonably accurate dates in real time can be approached through any of the various radiometric methods, where the age is determined through the radioactive decay of one element into another (**Fig. 1.8**). The principle is extremely simple to understand. If one has a given amount of "parent" isotope that was either created or buried at the time of deposition and in the same deposit a certain amount of the "daughter" substance into which it must change, and if one knows the rate at which the parent changes into the daughter, then it is a simple matter of determining how long it has been since the original parent substance was interred in the deposit. Let us consider an example using carbon.

Carbon exists in the atmosphere in two forms, stable C^{12} and its unstable isotope C^{14}. These forms of carbon are taken in by animals when they are alive. Although there is much less C^{14} around than C^{12}, there is a constant ratio between the two in a living animal. We also release each as carbon dioxide, so C^{14} never gets a chance to build up and begin to decay. However, upon death, the C^{14} that remains in bone begins to change to another element, N^{14}. (C^{14} does not change into stable carbon because when it decays a neutron changes to a proton, which changes its atomic number. Carbon has an atomic number of six, whereas the atomic number of nitrogen is seven.) The rate of this transition from C^{14} to N^{14} can be measured by a device similar to a Geiger counter that records the number of emissions of atomic particles per unit time. A rate of decay is expressed as a *half-life*. That is, it takes half an original amount of C^{14} 5370 years to decay to N^{14}. Half of the remaining C^{14} decays in an additional 5730 years and so forth. Because the ratio of C^{14} to C^{12} is so precise in living organisms, the age of a fossil is more often expressed in terms of the C^{14}/C^{12} ratio than the C^{14}/N^{14} ratio.

The radiocarbon method works, as do all other radiometric methods, because nothing affects the rate of atomic decay. This decay rate is not, for example, temperature dependent, as are so many chemical reactions. As long as the laws of physics operate and have operated in the universe (presumably back to around a billionth of a second after the Big Bang), this decay rate is, was, and will be unaffected. Radioactive decay truly represents the ticking of a universal foolproof clock. However, there are ways in which a date can be wrong. For example, one of the assumptions in radiometric dating is that the system under study has been completely closed since the original fossil or rock being dated was buried. That is, there has been no additional uptake or release of material. Clearly, if there has been any loss or contamination from another source, the date can be affected. We now know, for example, that various regional conditions (such as a volcanic upheaval) may affect the atmospheric ratio of C^{14}/C^{12}. But this is just mechanics; the method itself is not in question, and with care radiocarbon dating is highly reliable.

Different elemental isotopes have different half-lives, and the extent of the half-life and other conditions determine which isotope can be used for which circumstance. Radiocarbon, for example, can be used only on fossils or rocks dating back to about 50,000 years before present. This is because of the short half-life of C^{14}. After about 10 half-lives there is so little original radioactive carbon remaining that the signal is so weak it cannot be differentiated from the Earth's normal background radiation. So, if we are interested in dating rocks or fossils of greater age, in the millions or billions of years, we need to

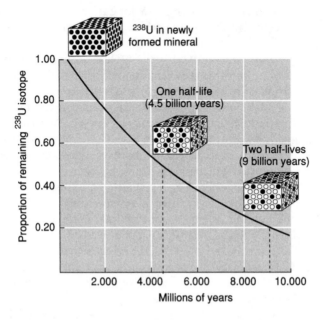

Figure 1.8 The decay of ^{238}U over time, assuming a half-life of about 4.5 billion years. Scientists can date rocks and bones based on the known half-life of a variety of isotopes. The half-life of ^{14}C, for example, is 5730 years, and accurate dates can be obtained for bones that are younger than about 50,000 years B.P. using this isotope.

use other radioactive systems. Fortunately, they exist. Uranium–thorium, uranium–lead, potassium–argon, and argon–argon are a few of the decay systems used in geochronology. The potassium–argon and argon–argon systems are especially useful for rocks and fossils in the range of the past hundred million years. The one limitation of these two methods is that they can be used only on sediments of volcanic origin, specifically on ashes and lava extruded from volcanoes. As soon as the radioactive potassium (K^{40}) in an ashfall hits the ground, it begins to decay to the gas argon. This gas is trapped in the crystal lattice of the volcanic ash and stays there until it is artificially released in the laboratory. The half-life of this K/Ar system is 1265 million years and can only be used on sediments that are at least a half million years in age, because it takes that much time for a reliably significant amount of argon gas to have accumulated. The K/Ar method cannot be used directly on fossils, but it can be used on ashes in which fossils are found or to bracket fossils in time. That is, if fossils are found in a stream deposit that sits between two ash beds and the lower bed yields a date of 3.5 million years and the upper 3.0 million years, we can conclude with some certainty that the age of the fossils lies between those dates.

Argon 40 also decays to a less unstable form of argon gas, Ar^{39}. This transition has been used in recent years with great success, and the literature suggests that it may be more accurate than K/Ar dating, especially on fossil localities less than one million years old.

Additional Methods

The heat from volcanoes is so intense that it melts minerals and creates glass-like rocks, such as obsidian. It is within these substances that we find the trapped argon from the K/Ar method and also "tracks" left by particles emitted from radioactive uranium. The analysis of these tracks to determine the age of the rocks since the last volcanic upheaval is known as *fission-track dating*. The amount of disruption (the tracks) can be read under a microscope and is directly proportional to the age of the rock. If the rock was heated after the first volcanic upheaval in which it was created, both the K/Ar and fission track methods can obviously become affected. Nevertheless, the fission-track method has been used to corroborate K/Ar dates and appears to be a valuable test system in its own right.

Thermoluminescence is a somewhat less reliable method based on the observation that when rocks are heated by internal radioactive decay, electrons are often trapped inside. When these rocks are superheated in the laboratory, the electrons are released and measured by the intensity of light they produce. If one can deduce the source and rate of original electron entrapment, then the intensity of light produced when the electrons are released in the laboratory can be used as an independent means of age determination. This method has the added benefit of being most accurate in the period of time between the upper limits of radiocarbon and the lower limits of K/Ar dating.

The last method to be discussed under real time dating is amino acid racemization. Amino acids, the building blocks of proteins, are found in animal bodies and may be preserved in fossil tissues, such as bone and shell. Many amino acids exist in two mirror image forms, the so-called L and D forms, or *isomers*. The L and D isomers differ in the way the amino acid molecule folds as a result of the angle of the atomic bonds that make it up. In living organisms, amino acids exist equally in the L and D form, but after death the L form changes, or *racemizes,* at a known rate to the D form. Theoretically then, the proportion of L and D isomers can, as with radiocarbon dating, give one an accurate age for a fossil. Unfortunately, racemization, unlike radioactive decay, is directly and easily affected by temperature. Heating speeds the process (giving errant ancient dates) and cooling slows it down (resulting in erroneous young dates). Because most fossil deposits are subjected to general seasonal heating and cooling and other less predictable influences, some variation in the racemization process can be expected since a fossil was originally deposited. This method is now used to augment other dating methods or as a general and tentative dating tool when other methods cannot be used.

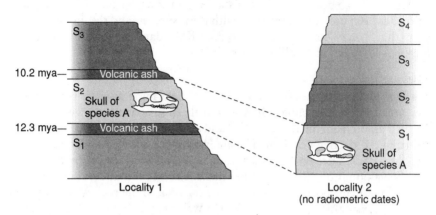

Figure 1.9 The principle of biostratigraphy. Because stratum 2 (S$_2$) at locality 1 and stratum 1 at locality 2 share the same species, they are considered to be roughly equivalent in age. This hypothesis would be further supported if the two strata had more species in common. In this example, dated volcanic ashes provide a range of real time for the age of the rock units.

Relative Dating Techniques

Relative dating could also be titled "sequence dating" or just "sequencing." It includes a group of methods that places a rock unit with fossils in a likely bracket of time, and the temporal refinement is dependent on the number of intervals that have been identified. Two general approaches are those we can lump as "biostratigraphy" (**Fig. 1.9**) and the other as "paleomagnetic correlation." Relative dating techniques begin with the initial premise, first proposed by the Danish physician and scientist Nicolaus Stensio (1638–1686), that older rocks normally underlay younger ones. This "law of superposition" seems simple, but it was a radical and important idea when first considered by Stensio in the seventeenth century, because it implied both a time sequence and natural explanation for geological features. In many places in North America and elsewhere, the geology of superposed rock units (strata) have been studied and described. Investigations of this kind fall under the geological discipline of *stratigraphy*. If we are fortunate, in some cases datable rock units, such as volcanic ashes, may be preserved, and an age for part of the sequence can be determined. This age then calibrates the entire sequence in time. However, in many rock exposures there are no datable sediments, so how can geologists and paleontologists know how old the rocks might be? In these cases, the age can be inferred to some extent by a principle known as *biostratigraphy*, in which the evolutionary stages of the fossils in the rocks provide some clues as to their age. The basic idea is as follows. Somewhere there are rocks with fossils that have been dated by radiometric means, and these fossils are in a particular stage in the evolution of the group to which they belong. In another geographical area we have rocks with fossils, but without sediments that can be radiometrically dated, and the evolutionary stage of the fossils in these latter rocks can be used to infer if the rocks are older or younger than the rocks with fossils that have been dated. In fact, this sequencing procedure can be done without any radiometric dates at all; the dates are only necessary to calibrate the sequence, to place it into real time.

Biostratigraphic methods are based on a single assumption, that evolution is not reversible. Sometimes known also as Dollo's Law, this precept allows us to chart the evolution of organisms, clades, and various ecological associations through time. We could also view this as a form of "Time's Arrow," the vector or directional nature of evolutionary change. Sequencing depends on "index fossils" or aggregations that indicate this vector. Such analyses generally begin with modern species and work backward, so we know what to look for in the fossil record. Variation among modern species is assessed, and it is often possible to work backward and, if the fossil record for a particular group is dense enough, develop a likely evolutionary scenario for the entire group. This scenario would place certain species at an earlier date than others. Now, if many of these scenarios could be determined, it would be possible to combine them to provide a strong argument for sequencing localities based on the fossil species found in them, even if we did not at first know how old they were. **Figure 1.10** presents a very simplified illustration of the chronological ranges of select rodent genera and species and the regional animal assemblages (local faunas) with which they are associated through the last five million years in the Meade Basin of southwestern Kansas. The vertical (superpositional) relationships among the assemblages was determined by careful field mapping over many years. Each local fauna was recovered from a single quarry or set of quarries at exactly the same horizon. Notice how the number of different generic or species lineages (e.g., *Geomys, Microtus pliocaenicus*) provides both additional refinement and reliability. Based only on a single genus we might consider the sequence highly tentative, but what would be the likelihood of a serious sequencing error with four or five such lineages? In the history of Kansas, a locality with *Microtus pliocaenicus* and *Sigmodon curtisi* is surely younger than one with *Sigmodon minor* and *Ogmodontomys*. The record of *M. pliocaenicus* in the Nash 72 local fauna represents the earliest securely dated presence of this species in North America. This vole is clearly an Asian immigrant and apparently came across the Bering land bridge between Siberia and Alaska during a period of low sea level. A series of volcanic ashes, dated by the K/Ar and Ar/Ar methods, gives us a few calibration points for some of the older sites (and, additionally, a narrow range of time for the *Microtus* immigration event), and radiocarbon dating adds reliability to the other end of the scale for some of the younger localities. Although we are fortunate to have a number of highly fossiliferous sites

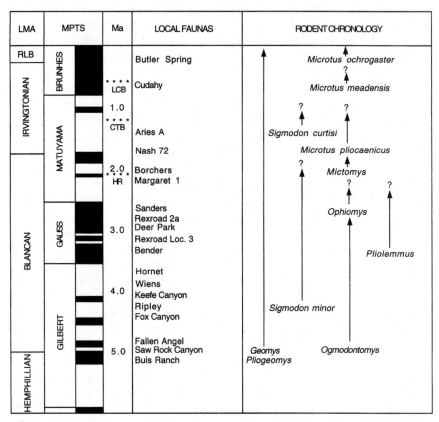

Figure 1.10 The sequence of fossil vertebrate assemblages (local faunas) from the Meade Basin of southwestern Kansas. The chronological ranges of some characteristic mammal genera (single names) and species (genus and species names) are provided. (Based on work of R.A. Martin, J. Honey, P. Peláez-Campomanes, G. Izett, C.W. Hibbard, and others.)

in Kansas and therefore the ability to create a useful biostratigraphic model, it is important to emphasize the model is also supported by stratigraphic (superpositional) information. One can rightly suggest that this kind of physical information must come first, or paleontologists run the risk of circular reasoning.

As noted in a previous section, the Earth's magnetic field has flip-flopped many times in the past. Investigators have assembled a fairly detailed history of these reversals for the past five million years. Each major period of normal and reversed magnetism is called a *chron*; minor intervals are called *subchrons*. Thus the Matuyama chron represents a reversal that lasted for more than 1.72 million years, from about 2.50 to 0.78 million years ago. The value of paleomagnetism is that it offers another powerful tool for faunal sequencing to the extent that paleomagnetic reversals are accurately dated. All the paleomagnetic events in Figure 1.10 are associated somewhere on Earth with radioisotopic dates. Now, if we have a fossil locality from Kansas that, on the basis of the small mammals, we can say likely lies between 1.5 and 2.0 million years before present, we can tentatively conclude that it resides more specifically between 1.77 and 1.95 million years ago if the sediments from this locality are normally magnetized, because the only recognized normal polarity during this period of the Matuyama reversed interval is a brief subchron known as the Olduvai event from 1.77 to 1.95 million years ago.

Suggested Readings

Condie, K. C. 1993. *Plate Tectonics and Crustal Evolution.* Pergamon, New York.

Darwin, F. (ed.). 1958. *The Autobiography of Charles Darwin and Selected Letters.* Dover, New York.

Eisley, L. 1961. *Darwin's Century.* Anchor Book, Doubleday & Co., New York.

Erickson, J. and E. A. Muller. 2001. *Plate Tectonics: Unraveling the Mysteries of the Earth.* Checkmark Books, New York.

Hawking, S. 1998. *A Brief History of Time.* Bantam Doubleday Dell, New York.

Hawking, S. 2001. *The Universe in a Nutshell.* Bantam Doubleday Dell, New York.

Kutter, G. S. 1987. *The Universe and Life.* Jones and Bartlett, Boston.

Tedford, R. H. 1970. Principles and practices of mammalian geochronology in North America. *Proceedings of the North American Paleontological Convention.* Part F:666–703.

Wilson, R. W. 1979. The cosmic microwave background radiation. *Science* 205:866–874.

2

The Origin and Evolution of Species

"Speciation is not evolution; it is a form of population fragmentation that severs reproductive continuity between ancestral and descendant populations and generally initiates evolution along new trajectories in the descendant."

Evolution is change. Darwin expressed it as "descent with modification." Basically, the environment selects from an overabundance of organisms that will leave their DNA to the next generation. This *natural selection,* as Darwin called it, culls out survivors from organisms that vary one from another in many ways. Individual variation in species is determined by a number of hereditary and developmental factors, and the way that selection acts on this variation determines both the fate and the evolutionary direction of species through time. Sometimes the beginning and end points in an evolutionary progression can be somewhat different biological entities. Darwin and others considered this kind of evolution within a single lineage as a legitimate avenue to the origin of new species. It was called *anagenesis* and is contrasted with another source of new species, *cladogenesis,* in which there is a splitting of one species lineage into at least two lineages. But can both processes be modes of speciation? After first examining the critical mechanism of evolution, natural selection, and the variation on which it acts, I present an argument that evolution and speciation are separate phenomena, with speciation acting as an important process that channels evolution into new and often unpredictable directions.

Before beginning this examination, it is necessary to differentiate between two levels, or scales, of evolutionary investigation and the expected results from each. *Microevolution* is a term that corresponds to changes that occur within populations of single species. It normally covers the process of natural selection, shifts in gene frequencies, and changes in populations that result over time. By contrast, *macroevolution* covers large scale and long-term

29

evolutionary patterns among many species, for instance, the origination, radiation, and eventual extinction of the dinosaurs; the evolution of the mammalian middle ear bones from the jaw and skull bones of ancestral reptiles; or the tendency of any *clade* (a group of related organisms and their ancestor) to evolve toward large or small size. One of the current controversies in evolutionary science is the extent to which microevolutionary mechanisms can explain macroevolutionary patterns.

What Is a Species?

This may seem fairly easy to answer, but the question has taken up a lot of space in important scientific journals and books and continues to be a hot topic even today. The casual observer sees mostly discrete entities in nature; a blue jay is clearly distinct from a robin, which is distinct from a nuthatch, and so forth. A king snake looks different from a pygmy rattlesnake, which looks different from a hog-nosed snake. But a herpetologist, a scientist who studies snakes, will tell you that a king snake, *Lampropeltis getulis,* in fact looks quite different depending on where you catch it. So do members of the species *Peromyscus maniculatus,* the prairie deer mouse. These species are *polymorphic,* meaning they display different forms. Forest populations of *P. maniculatus* have long feet and long tails, whereas prairie populations have shorter feet and shorter tails. For some reason that we do not fully comprehend, prairie populations sometimes will not breed with forest populations in certain regions, whereas they will in others. This phenomenon is also known in European crows and a few other species. It is called a *rosencreis* phenomenon, where the image of Christian rosary beads acts as a metaphor for populations of organisms that freely interbreed with each other at all but one place (where the clasp attaches the two ends of the necklace).

The most widely accepted species definition, introduced by the great evolutionary biologist Ernst Mayr, known as the *biological species concept,* speaks of species composed of individuals that actually or potentially interbreed with each other. By this definition, species are defined on the basis of their genetic integrity. They share DNA with each other but not with other species. So, theoretically, animals that do not interbreed with each other cannot belong to the same species. What to do then with the deer mice and crows? Some populations interbreed and others do not. What if, in addition, we discover that species change their morphology and size through geological time? Also, some populations that seem to be distinct species interbreed only in a small area of overlap, called a *hybrid zone.* Hybrid zones, at least during the lifetime of observers, appear to be stable. In some cases we are talking about an area that may only be a few hundred yards long and 50 yards wide. This is a widespread phenomenon, known in plants and in animals. So what does this say about Mayr's species definition? Can species really evolve as a unit if their boundaries are so indistinct? Do we need other species definitions? Some scientists respond strongly in the affirmative, and a number of new species definitions have appeared in recent years. I will not go into these here, in part because I am convinced that Mayr's definition is the most biologically meaningful. The most important points to remember are that evolution is a dynamic process and that populations within species often vary considerably from one another. In a very real sense, there are numerous experiments going on within species; each population exists in a slightly different environment from all others, and both natural selection and genetic accidents act to maintain these differences and continually create new ones. It is hardly surprising that the hereditary boundaries between species are sometimes fuzzy, and to the extent that mosaic evolution (see below) dominates in the history of a species, intermediate conditions are entirely expected. As we shall see, it is one of the amazing facts of the evolutionary record that despite these experiments, species do in fact seem to have distinct births (originations or "speciation events"), lifespans, and deaths (extinctions).

Variation and Evolution in Populations

Natural Selection

The forests of England are home to a very famous insect, the English peppered moth, *Biston betularia*. It is famous because it has become a classic example in textbooks of Charles Darwin's fundamental evolutionary principle of *natural selection*. Other writers before Darwin speculated about evolution, but only Darwin and his contemporary, Alfred Russell Wallace, a plant biogeographer, provided a mechanism whereby such change could take place. Darwin's elegant and thorough discussions of evolution in the *Origin of Species,* published in 1859, eclipse Russell's writings, and credit is properly given to Darwin for the development of the modern theory of evolution and natural selection. The history of the peppered moth in England during the late nineteenth century provides an example of how natural selection can work to change an organism's appearance.

Biston betularia exists with two color morphs, light and dark (melanic). The light form is actually a mosaic of light and dark, hence the "peppered" name. This is not uncommon in nature; many species demonstrate a range of external color, sometimes in the same geographical area. In this case there is no gradation, just light and dark specimens. Before about 1848, the light morph was the most common. After 1900, however, the melanic morph replaced the light one throughout England in the vicinity of large cities. Scientists traced the shift to the Industrial Revolution and its effects. The light color

Figure. 2.1 Natural selection by birds on the English peppered moth since the Industrial Revolution may be responsible for the shift from the light to the dark color morph near cities, although recently the scientific rigor of some of the original experiments has been questioned. (Science VU/Visuals Unlimited.)

pattern of peppered moths is used as camouflage on trees where they seek refuge when not flying. In particular, they blend in against a mottled pattern of the trunk caused by the presence of a light-colored lichen (**Fig. 2.1**). Their primary predators are birds, and when the moths remain still on a tree they normally blend in quite well. However, after the Industrial Revolution got into full swing, the moths' environment changed. Soot from the many coal furnaces blanketed the forests, turning the once beautiful mottled trunks a dingy gray as most of the lichens died in the polluted areas around the cities. Studies have shown that the light moths were soon at a disadvantage and became easier for the birds to spot and to consume. The once rare dark moths, however, prospered and increased their numbers—natural selection in action. Where forests are unpolluted and the lichens survive, the light morph dominates. H. B. D. Kettlewell was able to show that birds were indeed able to discriminate between the morphs in a series of experiments where he transplanted light and dark moths to areas where they would be more conspicuous. The colors of the moths can also be viewed as *adaptations* that allow the moths to survive under particular ecological circumstances. Some organismal characteristics seem to be very clearly related to their survival (and are thus properly viewed as adaptations), whereas others may not be and are then said to be *adaptively neutral.*

Darwin's mechanism of natural selection is based on the premise that organisms in nature produce more offspring than can normally be supported by available resources, and this "superfecundity," as it is sometimes called, leads to a selection of the fittest to survive. Darwin got this idea, in part, from a well-known article by Thomas Malthus, an English cleric of the mid-eighteenth century who expressed the idea in terms of a theoretical over-population by humans. Fitness is measured by reproductive success. Those individuals that survive and reproduce maximally, thus passing their DNA to the next generation at an elevated rate for the species, are deemed most fit by definition. I mention the rate of reproduction here because it is not sufficient merely to reproduce. All other things being equal, a family of organisms that inherently produces more offspring per generation than another will replace the latter. That is, if all Smiths in the phonebook are truly related and if every Smith female gives birth to five children for every one or two that females by every other name produce, at some time in the future the world will likely be populated only by Smiths.

Natural selection also shows us the way by which morphology (anatomy) can change through time. Darwin knew that by selective breeding one can change a breed or plant variety (**Fig. 2.2**). That is how we have large sweet ears of corn and cockapoos. This "artificial" selection is fundamentally exactly what goes on in nature, except that in nature the vagaries of the abiotic and biotic environment act as the selecting agents. It is a sad testimony to the speed and efficiency of information transfer in the late 1800s that Darwin did not know of the critical experiments in heredity being performed by an Austrian monk by the name of Gregor Mendel. Darwin died without a clear understanding of the mechanisms of inheritance despite the fact that the father of the modern field of genetics was a European contemporary. Think of the progress that could have been made if Mendel and Darwin had been able to communicate on the Web. We now know that only one gene with two forms (alleles) controls the light and dark color of peppered moths. The melanic gene C is dominant to the light gene, c. So moths with the genetic combination CC and Cc are dark; only those with cc are light. This shows how powerful the selective force of predation by birds must be in unpolluted forests to keep the light form dominant. Simple math shows that under circumstances of random mating and no selection, the melanic form would normally be produced at a rate of 3:1 relative to the light colored morph. Which leads us to the next topic, the Hardy–Weinberg equilibrium.

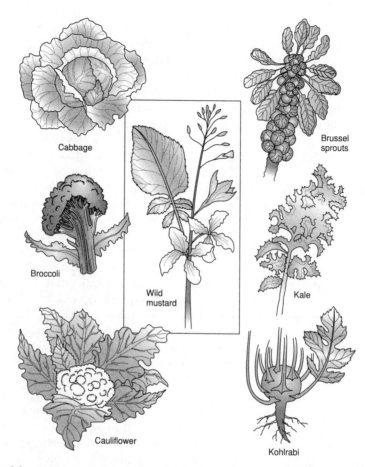

Figure 2.2 Variation in edible vegetables all derived from the wild mustard. The different varieties were produced through artificial selection, a speeded-up and human-directed version of natural selection, the mechanism proposed by Charles Darwin for evolution of species (lineages) in nature. (Adapted from N. Campbell, L. Mitchell, and J. Reece, 2000. *Biology: Concepts and Connections,* 3rd ed. San Francisco: Benjamin Cummings. p. 264.)

Two biologists named Hardy and Weinberg independently came up with a mathematical proof showing that in the absence of natural selection and other potential influences and after one generation of random mating, gene frequencies in a population would remain unchanged in subsequent generations. This also means that organismal appearance would not change. This is an important discovery, because it makes it easier to then determine the agents for change or evolution. Studies have confirmed many features in organismal populations that conform to the Hardy–Weinberg equilibrium, for instance, some of the human ABO blood groups. To a geneticist, evolution can be defined as the extent of shifts in gene frequencies. It is certainly true that one cannot have a change in organismal appearance without a change in gene frequencies, but we must also remember that natural selection acts on the animal as a whole, not on its genes.

Agents of Selection

Nature selects. But nature has many weapons at her disposal. For instance, the innate reproductive rate of a population (abbreviated as r) may determine an evolutionary outcome, as we saw above with the Smiths cranking out kids faster than anybody else. This is a form of *passive competition,* and there are many others in this general category. Animals do not need to bloody themselves against each other to compete, though that does, in fact, occur as well. Male lions that have ousted a rival often kill the rival's cubs, thus eliminating its genes. Mice of the genus *Sigmodon* will attack and kill pups of other species in its daily range. We may label this as *active* or *interference competition,* resulting directly in the death or banishment of a rival individual. These two examples also serve to differentiate two additional categories of competition: *intraspecific* and *interspecific.* Intraspecific competition is limited to individuals within the same species. Interspecific competition applies to competition between members of different species. Both kinds of competition occur in nature, and both are therefore forces that can channel a species into various vectors of change through time. Organisms compete for all the natural resources they need to survive and reproduce: space, food, shelter, mates, water, heat, and so on. The list is almost endless and differs from species to species and even among different populations of the same species. And as individuals compete for these resources, the resources themselves can act as selective forces. Heat can provide the needed internal temperature levels for lizards to forage and can also provide appropriate growing conditions for the green plants that form the basis of most ecosystems. However, in the absence of adequate rainfall, lush environments can turn to deserts. When they do, wholesale slaughter occurs, and few individuals from any species survive, which is natural selection in the extreme. This is likely the reason so many desert plants have deep roots; only those with the deepest roots could reach the meager water that remains available. So here we see that Mother Nature herself, the abiotic environment, does the selecting, and the population biology of a species may be of little consequence. (Clearly, a large population has a better chance of surviving such harsh conditions than a small one, because there is likely more variability in a large population and therefore the probability will be greater that a few individuals will possess whatever characteristics are necessary to make it through the calamity. But the fossil record is quite literally strewn with the bodies of extinct species. No matter how well adapted a species can become, there are some selective agents so powerful that in the long run they can also work to cause the demise of an entire species. This is an interesting and apparently contradictory influence, and we examine it a bit later under the general topic of extinction.)

Animals generally follow the plants, not vice versa. So the presence of particular plant communities will determine, or select, the specific animal aggregations found there. There is considerable evidence that plant communities have changed dramatically during the past. In fact, modern plant communities in most areas of the world are less than 10,000 years old. As plant communities change in composition, animal species in them are replaced by other species better suited to the new plant associations. For example, during the early part of the Cenozoic Era, about 60 million years ago, primitive monkeys were common members of ecosystems in western Europe and western North America. The environment was quite different then; subtropical to tropical conditions and forests prevailed. Global weather patterns changed throughout the Cenozoic, resulting in a restriction of tropical environments to equatorial regions. To the north, more temperate conditions developed, with distinct cold winter periods and a switch to the forests that now dominate those regions, mostly deciduous hardwoods and coniferous evergreens. These shifts in climate and plant communities were eventually accompanied by a change in animal associations. In some cases variation within a species was sufficient that it could change to meet the new challenge. In others the variability was insufficient, and the latter species became extinct, giving way to new and better adapted immigrant species. Such patterns of replacement have been documented in almost every animal group throughout the Cenozoic. It has become clear that there is an evolutionary benefit for species to be as variable as possible within the limits of the environment where they reside. This variability serves as a cushion when conditions change.

Population Responses to Selection

When a selective agent acts on a population, only three outcomes are possible: the population does not change, the population changes in a single direction, or the population splits, representing at least one and possibly two new evolutionary trajectories. These actions are termed, respectively, *stabilizing selection, directional selection,* and *disruptive selection* (**Fig. 2.3**). It is one of the amazing and predictable aspects of nature that almost every quantity one measures about an organism displays a *normal distribution*. This is a statistical term meaning simply that most individuals in a population are similar for most characteristics (e.g., body weight or height), with a few outliers, or extremes, in both directions. Another statistical term, the *mean* or average, is a computation expressing this central tendency. The mean is just the total of the individual measurements taken in a population divided by the

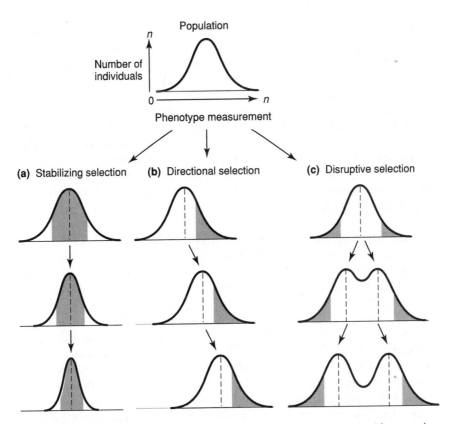

Figure 2.3 The three possible outcomes of natural selection on a characteristic with a normal (bell-shaped) frequency distribution. In the case of (a) stabilizing selection, natural selection works to preserve the average value. In (b) directional selection, there is active selection against one extreme (tail) of the distribution, and the mean, therefore, shifts through time. In (c) disruptive selection, selection works against average individuals, and two distinct morphs can theoretically result.

population size. For example, average weight for males in a human sample might be 165 pounds, with a range from 105 to 325 pounds. When a normal distribution is expressed graphically, a so-called bell curve results. If we think in terms of these curves, we can develop a visual image of the three types of selection. Stabilizing selection acts to keep a population where it is. An example of this is birthweight of human fetuses. The optimal birthweight for a human is about 8 pounds. Infants born much below and much above this weight often die. Selection has optimized the weight that provides the most likelihood of survival and normal development. The mean stays stable.

Directional selection occurs when a proportion of a population, expressing a distinct part of its range of variation, is removed, thus shifting the remaining population mean either up or down. The example discussed earlier with the English peppered moth is a perfect expression of directional selection. Directional selection appears to be the most common form that enacts change in populations. It is likely to be responsible for virtually every evolutionary trend in species lineages, many of which are discussed in detail in the case history chapters in this book.

Disruptive selection is mostly a theoretical concept, although this may be more due to the fact that it is difficult to document than to its rarity. Under this concept, a selective agent splits the initial population into two, each with its own mean for one or more features. This is a likely scenario for the process of *sympatric speciation,* where a descendant species arises from within the geographical range of its ancestor. This topic is considered in more detail in a later section of this chapter.

Sexual Selection and the Origin of Anatomical Novelties

The anatomy of some animals is gaudy almost to the point of attaining the bizarre. The huge antlers on moose and reindeer, the dazzling plumage of the Bird-of-Paradise and the peacock, the wavy and elongated tails of many tropical fish, and the extraordinary coloration of males of many species are but a few of the more obvious examples. Sexual dimorphism (the tendency for males and females to be of different size) is also common in nature, even in humans. Darwin's explanation for these differences is rooted in an extension of natural selection he called *sexual selection.* It is not a different mechanism from natural selection, but the selective agent is always either the male or female of the species. If the primary "goal" of the evolutionary process is to maximize fitness as measured by reproductive success, then it naturally follows that any changes in organisms that would enhance the reproductive process would have positive selective value. A male that mates more frequently with more females than other males will leave more offspring, and his genes will be represented more often in following generations. Similarly, a male with features preferred by more females will mate more often. Two forms of sexual selection are recognized by biologists: *intrasexual* selection, mostly among males, and *intersexual* selection, most often expressed by females choosing males based on a variety of attributes.

Males of many species are much larger than would be expected if the only factor determining size was food availability (energy capture); they may be in excess of twice the body mass of females. Experiments in many animal taxa have confirmed that larger size is most often related, in one way or another, with mating. In reptiles and mammals, for example, larger size is often seen in animals that set up and defend mating territories (e.g., marine iguanas, deer and their relatives, lions). Larger more aggressive males mate more frequently and thus leave more offspring. Intrasexual selection can lead to the appearance and development of many traits besides just size, including horns and antlers, spikes, and body armor (of course, carnivory may play a part in the evolution of protective features in herbivores, so there may be some overlap of selective agents). Intrasexual selection also leads to what biologists call "sperm competition"; after all, mating is not the end of the game. The sperm have to fertilize the eggs. Consequently, we see that a variety of strategies has appeared to facilitate this process. Males in some species mate copiously, with many different females. Others maximize their effort by producing relatively huge ejaculates with many sperm, prolonging ejaculation, applying pheromones (sex hormones) that reduce female attractiveness, or, as shown by J. K. Waage in damselflies, scooping out sperm left by previous males. Male red-tailed and brush-tailed phascogales, small carnivorous marsupials from western Australia, use up so much energy during mating that they die soon after, probably because their immune systems have become compromised. Infanticide, practiced by lions, is also an example of intrasexual selection. After displacing a rival male, the successor may kill the other male's young. This accomplishes two evolutionary objectives: ensuring the other male's genes do not proliferate and bringing the adult female into estrus more quickly, thus initiating passage of the successor's genes more quickly. The long neck of giraffes may be a result, say Robert Simmons and Lue Scheepers, not of selection for better foraging ability on higher growing plants, but as a means to increase the centrifugal force of the head as a weapon during aggressive encounters. Males use their neck and head during fights over mates and may even kill one another in the process.

Females of most animal species play a much more dominant role in selecting male attributes than was thought for many years. Experiments with many species have confirmed that females are often very particular with which males they will mate. Some of the more intriguing experiments have been done with tail length in fish and birds, mating calls in amphibians and birds, and eyestalks in stalk-eyed flies. In each case manipulations of features demonstrated that a statistically significant percentage of the females chose to mate with males with enhanced features. Why do the females respond this way? It turns out that the genes predisposing females to choose exaggerated traits may be there before they appear in the males. This possibility is demonstrated by experiments conducted with finches by Nancy Burley and with platyfish by Alexandra Basolo as reported by Douglas Futuyma in his book, *Evolutionary Biology*. Grassfinches do not have feather crests on their heads, as do North American cardinals, for example. Nevertheless, when fitted with artificial crests, female finches preferred the crested males to typical ones (**Fig. 2.4**). Similarly, female platyfish preferentially chose males with artificial plastic swords attached to their tails even though male platyfish do not have swords. It looks like sexual selection has a lot of power and may help explain most

Figure 2.4 Female and male grassfinches fitted with artificial feather crests. Although grass-finches do not have crests, females prefer males with crests over those without crests, suggesting that the gene for such preference is present in females prior to the evolution of the structure. (Photograph by K. Klayman, courtesy of N.T. Burley.)

of the exaggerated features we see in the sexes of many species. (I do not indicate just the males here, as there are species in which the females are much larger than the males; black widow spiders, for example.)

Sources of Variation

Natural selection acts upon biological properties of organisms, their size and shape, color, behavior, intelligence, physiological responses, and many other features. These things are individually referred to by geneticists as *phenotypes*. Natural selection works on these phenotypes, but they, in turn, are controlled by an underlying genetic program known as the *genotype*. That is, for every organismal property, or characteristic, there is a gene or gene combination that determines its expression. Some phenotypes are expressed in a binary fashion, either on or off, whereas others, such as intelligence and height in humans, are the result of the blended interaction of many genes. The genetic program, or code, is translated into all the parts that are an organism by the process of development, or *ontogeny*. Human development begins with a fertilized egg and lasts for years, including embryonic, infant, and adolescent stages. At all stages there is a general similarity of form, but this is not the case for all organisms. Many insects, for example, demonstrate a complete metamorphosis, in which there are three stages—egg, larva, and adult-none of which resembles each other. It is one of the scientific wonders that a worm-like creature can be reorganized to become a delicate and beautiful butterfly. Natural selection operates at all stages during the lifetime of an organism; embryos, like the adults they will become, must be adapted to their environments.

There is an almost infinite variety of individual features in organisms; no two humans are alike, for example, not even so-called identical twins. But what is the source of this incredible variation, and why is it there? As we shall see, hereditary mechanisms act to ensure a large pool of variation from which Nature can select those most suitable to survive. Superposed on this is the influence of the environment; some genetic/developmental systems are so flexible that specific habitat signals can lead the developmental process down more than one pathway.

The Hereditary Program

Many excellent texts explain the basic principles and underlying chemistry of heredity. In this section I endeavor only to provide a summary of the mechanisms by which hereditary systems function to produce the variation in organisms upon which natural selection exerts its influence.

Genes are segments of a DNA molecule. DNA is the stuff that makes up chromosomes. A place on a chromosome where a gene is found is a *locus*. Different forms of a gene at one locus are *alleles*. Every cell of the human body contains a full set of chromosomes, in our case 23 pairs, 46 in total. This is known as the 2N or *diploid* number. If we had the technology, we could clone you from a cell of the lining of your small intestine or your mouth (because scientists recently cloned a lamb, the technology may actually be here already). The complete blueprint for a human is in every human cell. During the processes by which eggs and sperm are produced in sexually reproducing organisms, the normal cellular component of chromosomes becomes reduced by half, resulting in the N or *haploid* number. The process by which this occurs is called *meiosis*, or reduction division (**Fig. 2.5a**). Specifically, it is known as oogenesis in producing eggs and as spermatogenesis in producing sperm.

During one of the final stages of meiosis, all the chromosomes line up at the middle of the primordial sex cell and before they go to their respective daughter cells they exchange some genes, in a process known as *crossing over*. Crossing over of a small percentage of genes takes place every time reduction division occurs. So there is always a shuffling around of genetic material through actual physical exchange of genes. Geneticists can create "maps" of genes on chromosomes from the frequency of crossing over that occurs at a particular locus. The farther the genes are from the center of the chromosome, the more likely they are to cross over.

Once the crossover genes have settled into their chromosomes, then new genetic combinations occur simply as a result of the rearrangement of chromosomes that also takes place during meiosis. It is entirely a matter of chance

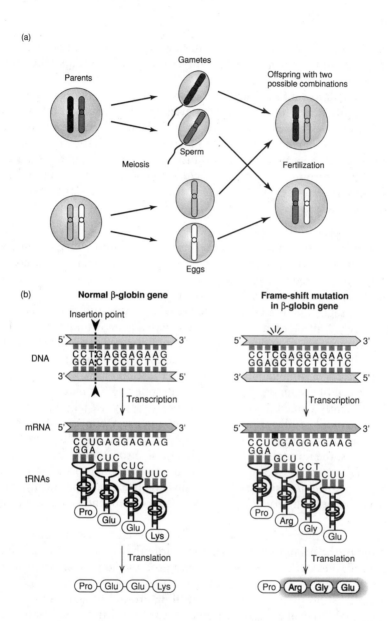

Figure 2.5 Gene recombinations during meiosis and fertilization (mating) (a) and mutations (b) provide the raw source of genetic (genotypic) variation upon which natural selection operates. The "frame-shift" point mutation in (b), producing a slightly different molecular structure of the b-globin molecule (a blood protein), is caused by the random insertion of a cytosine-guanine pair of nitrogenous bases in part of a DNA chain. Other point mutations include duplications and deletions of base pairs.

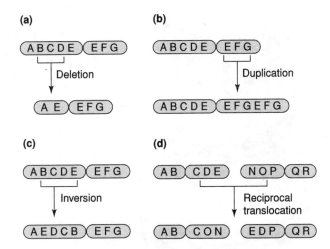

Figure 2.6 Chromosomes can mutate by (a) deletion, (b) duplication of a part, (c) inversion, or (d) translocation. Translocation may be either "reciprocal," in which the two chromosomes exchange equal lengths of DNA, or "non-reciprocal," in which one chromosome gains more than the other. In addition, whole chromosomes may fuse, and whole chromosomes (or the whole genome) may duplicate. (Adapted from M. Ridley, 1993. *Evolution.* Oxford: Blackwell Science. p. 74.)

which of a given pair of chromosomes lines up at which side at the center of a primordial sex cell before it splits to form the haploid gametes. Consider the number of possible combinations among 23 pairs of chromosomes, each with about 5000 genes, some of which have also crossed over. It is hardly any wonder that no two humans look the same, even without the next category of variation: *mutations.*

A mutation is any unprogrammed change in gene or chromosome structure and number. Mutations may be small changes in base pairs at a single locus, a *point mutation* (**Fig. 2.5b**), or something more dramatic, such as a doubling of the normal chromosome number, called *polyploidy* . Polyploidy is rare in animals but common in plants. All combinations in between are also known. A chromosome may be absent in one gamete and double in another. This can result in either a chromosomal deletion or a tripling in a fertilized egg. Such large changes are often detrimental. Trisomy 21, or having three number 21 chromosomes in humans, results in Down syndrome. These larger changes do not seem to add to normal healthy variation in organisms, but somewhat smaller changes may. Parts of chromosomes may become detached and reattach to other chromosomes (**Fig. 2.6**). As long as they are duplicated, these "jumping genes," as they are called, may allow new genetic combinations.

Mutations are rare, but given the large number of genes in a human, probably from one to two point mutations occur every time a sperm or egg is produced. These mutations contribute to overall genetic variation, or genetic polymorphism, and some have been shown to be adaptive. One of the classic experimental organisms in genetic research is a genus of fruit fly, *Drosophila*. They are the nuisance flies that swarm up when you forget to throw away an old piece of fruit. These tiny flies have a tremendous reproductive output, in

Figure 2.7 Variation, or polymorphism, in color pattern in a moth, *Panaxia dominula*. Each pattern is caused by a different combination of alleles, or forms of a gene. (From D.J. Futuyma, 1998. *Evolutionary Biology,* 3rd ed. Sunderland, MA: Sinauer Associates. p. 234.)

part because they can be born, metamorphose from worm-like larvae, and can be ready to mate and lay eggs in a period of 14 days. One of the species, *Drosophila pseudoobscura,* is widely distributed in the United States. Theodosius Dobzhansky, a founding father of modern genetics, demonstrated many years ago that the same genes were not always found in the same place on *D. pseudoobscura* chromosomes; that is, they had been rearranged due to various mutations that had taken place in the past. He also found that these patterns (which could actually be seen under a microscope in the giant salivary gland chromosomes of *Drosophila* as dark bands) varied with latitude. At each specific geographical location, there was a different frequency of the various genetic patterns. The proportions of the different genetic arrangements also varied with seasons at a single geographical location, suggesting some local adaptive benefit depending on environmental conditions. Many other examples of individual variability could be cited (**Fig. 2.7,** and see Fig. 2.11a). Snake species are notorious for their variation in color and banding patterns, as are snails. In England, the snail genus *Cepaea* has been studied for many generations, and it is clear that a complex combination of avian predation and temperature extremes act to select those individual snails best fit to survive. Although there are a few parthenogenetic vertebrate species (some lizards) consisting only of females, their rarity proves the rule that sexual reproduction in vertebrates and the variation it produces is a better evolutionary strategy.

Figure 2.8 The axolotl, a permanently neotenic salamander. Notice the feathery external gills in this adult individual. (Copyright Stephen Dalton/Photo Researchers, Inc.)

Developmental Influences

In one way or another, all anatomical change in or between species is a result of changes in developmental growth vectors. Sometimes the embryonic stages are extended, and development is retarded. This results in embryonic features showing up in adults. The name for this general process is *paedomorphosis*. Any change in the timing of developmental processes, regardless of the direction or mechanism, is known as *heterochrony*. A classic example of a paedomorphic feature is the flat face in humans. The face in nonhuman primates is extended; in baboons it forms a muzzle as in dogs. Apparently, during human evolution the genes that control skull growth began to code for smaller jaws and larger brains. The result was an expansion of the cranium at the "expense" of the face, sort of like Newton's First Law of Motion: For every action there is an equal and opposite reaction. Another example is the axolotl, a Mexican salamander that retains juvenile features such as gills and a laterally compressed tail into adulthood (**Fig. 2.8**). In most salamanders an aquatic larval stage precedes metamorphosis to a terrestrial form in which features useful in the aquatic mode, such as gills, are lost. The axolotl remains aquatic and maintains its gills and general aquatic form while its gonads mature.

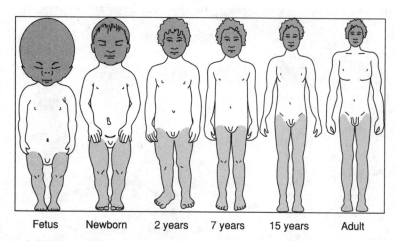

Fetus   Newborn   2 years   7 years   15 years   Adult

Figure 2-9 Relative growth of the human body illustrating the principle of allometry. The shaded areas indicate that following birth the head grows proportionately slower than the rest of the body, especially the legs. (Based on F. Beck, D.B. Moffat, and J.B. Lloyd, 1973. *Human Embryology and Genetics.* Oxford: Blackwell. p. 501.)

In another form of ontogenetic change, modifications are tacked on at the end of development. We call this *peramorphosis*. This is a common pattern seen in the evolution of many structures, such as the complexity gained in horns and antlers in a variety of artiodactyl groups such as African bovids and North American cervids. Over thousands of years slight changes in growth patterns can lead to new structural arrangements through changes in *allometric growth* relationships. The prefix *allo* means different and *metric* means proportion or measure. So allometric growth simply means different rates of growth in the same or related structures (**Fig. 2.9**). For instance, if your face grows slower than your braincase as your head grows during development, you end up looking like a human. If your face grows faster than your cranium, you end up looking like something from the movie *An American Werewolf in London*. You also end up pretty stupid. Many years ago the zoologist D'Arcy Thompson showed how slight changes in growth vectors could lead from one body form to another in a number of vertebrate groups (**Fig. 2.10**), and we are just now developing the computer modeling techniques and genetic comprehension to be able to investigate this process seriously. The genetic code must obviously be there, but because development represents the phenotypic expression of that code (the organism realized), it is important that we understand how the full process operates.

In recent years it has been possible to identify the structure and function of specific genes and their positions on particular chromosomes. As a result, geneticists have identified a series of genes, called *homeotic* genes, that con-

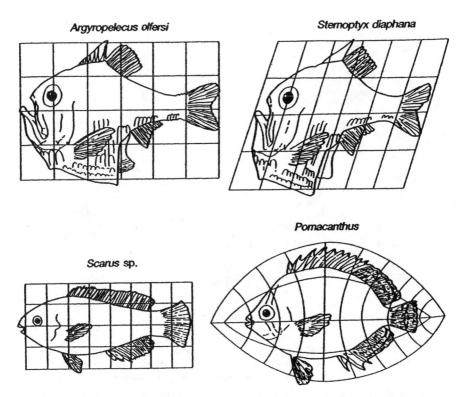

Figure 2.10 Different body forms in related fishes, showing how simple changes in geometric proportions (allometry) can explain the seemingly extreme differences in morphology. First noted by the English biologist D'Arcy Thompson, we would now explain these differences as due to varying proportional contributions of enzymes and growth factors during development, perhaps controlled only by one or a few genes. (From M. Ridley, 1993. *Evolution.* Oxford: Blackwell. p. 548.)

trol large scale developmental processes in many organisms. For a long time we have known that some genes are *regulatory* in function; that is, they do not code for specific structures. Instead, they code for the rate and timing of developmental events. The allometric modifications mentioned above could be affected by these kinds of genes. But only recently have geneticists come to realize distinct regions of homeotic genes on specific chromosomes, called *Hox* genes, that are responsible for the basic developmental patterns of animals. Invertebrate animals have one *Hox* cluster on one chromosome, whereas vertebrate animals may have up to four *Hox* clusters on different chromosomes. This difference appears to be related to different levels of complexity, and it is thought that the increased numbers came about by mutations resulting in duplications of *Hox* clusters. We now also know that the linear arrange-

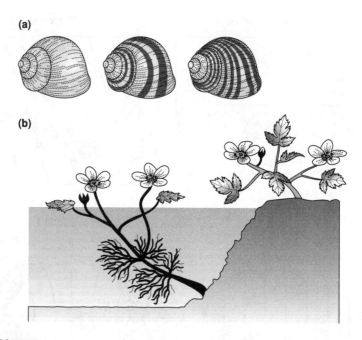

Figure 2.11 Variation in organisms may be under strict genetic control, as is the banding pattern in the English woodland snail *Cepaea nemoralis* (a) or it may be somewhat variable (an ecophenotype), as in leaf form of the water crowfoot, *Ranunculus flabellaris* (b) depending upon habitat.

ment of the *Hox* genes on a chromosome corresponds to their exact linear expression on an animal during development.

This knowledge helps us to understand certain well-known developmental phenomena. For example, all vertebrate embryos look very similar at certain points in their development. The time during which they are maximally similar (all have gill slits, large eyes, a tail, segmentation, etc.) is called the *phylotypic* stage. The modern interpretation of this stage is that it is determined very precisely among all vertebrates by a common *Hox* cluster. This conservation of form is thought to be beneficial in that it is during this time that basic symmetry and growth patterns are established that ensure normal development. After this stage, other genes kick in that take the animal to its particular final form. Studies of the inheritance, form, and function of homeotic genes may lead to a clearer understanding of the evolution of animal structure, such as the appearance of limbs in tetrapods (four-footed animals; see Chapter 10).

Environmental Influences

Some species have a developmental programming that is amazingly flexible. As noted earlier, the genotype is translated into the phenotype via development, or ontogeny. In some species ontogenetic pathways are quite expansive, and the environment can actually influence the adult morphology of an organism. For instance, in the Virginia oyster, *Crassostrea virginica,* shell shape is determined to some extent by substrate and current. In fast moving water and on a clean hard bottom, the shells will be long and heavily ribbed, whereas on a muddy bottom and in a slow current, the shells may be smooth and wide. If transferred between habitats, the shells will change to conform to the particular environmental conditions. In an aquatic plant *Ranunculus flabellaris,* the water crowfoot, leaf shape depends on whether or not the plant is submerged (**Fig. 2.11**). If it is, the leaf will be deeply notched; if not, it will demonstrate a smooth perimeter. These morphotypes are often referred to as *habitat forms,* or *ecophenotypes,* to separate them from morphologies that are controlled by different underlying genetic combinations. When studying fossil specimens, it is sometimes unclear if larger animals occurring in progressively older or younger deposits are the result of natural selection favoring animals with a distinct genotype or instead are the result of better forage or longer growing seasons acting on animals with the same genotype. Regardless of the genetic system controlling them, habitat forms represent a type of variability that can be responsive to environmental change and, therefore, under certain conditions, allow a species to survive.

Another expression of an unprogrammed modification in phenotypic expression is referred to as the *maternal effect.* Robert Bader showed many years ago that the spacing of teeth in adult house mice is determined to a certain extent by suckling when young. Because teeth are intimately involved in energy processing, their position in the jaw can be critical. There are other maternal effects; the quality of mother's milk can affect phenotypic expression in young mammals, for example, in a host of characters.

In summary, an underlying genetic program, enriched by mutation and recombination (Fig. 2.5b) and expressed through development, creates an almost unlimited source of variation in organisms. This variation is the stuff on which natural selection operates to ensure that the best fit individuals within and among species survive at any point in time.

Neutral Characters and Genetic Drift

The agents of natural selection theoretically channel evolution in directions that are, at least in the short run, *adaptive* for a given species. That is, the features of organisms satisfy specific requirements and would not be present otherwise. But the notion that all features of organisms, be they behavioral, physical, or physiological, must be adaptive has been seriously challenged in recent years. In part, this is because of the very variability that characterizes most species. Is it possible that every slight variant is critically adaptive to a specific environment? Or is there a range of variation that is produced, all of

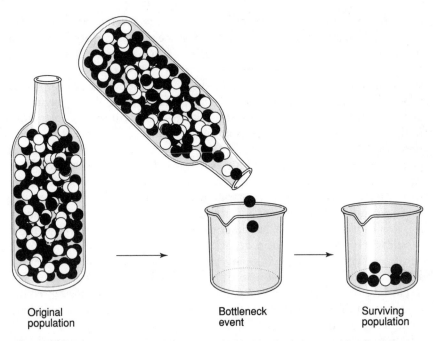

Original   Bottleneck   Surviving
population   event   population

Figure 2.12 A genetic bottleneck effect occurs when an ecological catastrophe eliminates a large proportion of individuals from a species, and only a small part of the original genetic variation remains in the survivors. (Adapted from N. Campbell, L. Mitchell, and J. Reece, 2000. *Biology: Concepts and Connections*, 3rd ed. San Francisco: Benjamin Cummings. p. 270.)

which is within tolerable limits, simply *because it is not maladaptive?* This may seem like double-talk: If a feature is not maladaptive, then it must be adaptive. To the extent that a human nose is necessary for breathing, yes; but there are many sizes and shapes of noses, and some are particularly more common in certain human populations. Are we to believe that all forms of the human nose are adaptive to the place where the humans arose initially? Or is it possible that once the nose had evolved with its basic function, it was then free to vary in many different ways, as long as the nasal canals were sufficiently large for adequate gaseous exchange? The many shapes of human noses would not, under this interpretation, be adaptive. They would be *adaptively neutral*. It is difficult to believe that the yellow cheeks of the yellow-cheek vole, the many color variants within certain African cichlid fish and butterfly populations, and the numerous dental patterns in voles have any particular adaptive value. That is not to say they could not be important in the future, but for the moment they seem to be "allowed" because they are functional and do not interfere with the reproductive success of the species.

Adaptively neutral features can also arise from a phenomenon known as *genetic drift*. Each population in a species has a set amount of variation determined by the way in which genes are shared among individuals. As noted earlier, many characters in a population express a normal distribution. However, under some circumstances that are very likely to occur in nature, this variation can become quickly limited by a mechanism that is not selective. Consider species of organisms on islands. Most islands have populations of animals that are related to others on either nearby islands or the closest mainland. Invariably, however, these island populations differ significantly from their geographically disjunct relatives. How did these differences arise? Are they adaptive? Many recent studies of variability of organisms on islands lead one to believe that their features are the result of a random drift process known as founder effect. Founder effect is the establishment of a new population by one or a few colonizing individuals. Obviously if one individual, it must be a gravid female. Animals get to islands by swimming or rafting on plant debris during storms. These new colonizers may have only a small percentage of the genetic variation of their ancestral population. Once on the island, and then given the genetics of interbreeding in small populations, certain features can become eliminated and others fixed. There are many examples, mostly of pathologies, conferred on modern human populations by a few settlers in years past. The presence of Huntington disease in Australia and Mauritius can be traced to specific founding families.

Another form of neutral drift can occur when a species experiences a severe reduction in population size, perhaps due to a natural calamity or disease, resulting in a limited representation of the original genetic variability (**Fig. 2.12**). Such genetic "bottlenecks" are difficult to identify but must surely have occurred in the past. Genetic variability in the cheetah in Africa is sometimes cited as likely the result of a population crisis.

Mosaic Evolution

To a population geneticist the expression of natural selection on individuals in populations over time resulting in local adaptations is *mosaic evolution*. In this model, populations within a species can be expected to demonstrate regional variation consistent with regional selective biotic and abiotic forces. If selective influences change in one area of a species range, populations in that area may likewise change, whereas populations elsewhere may not. In this fashion, it is possible for a species to demonstrate both derived and underived character states simultaneously, depending on the population sampled. So, in a sense each species has embedded in it many evolutionary experiments, some of which may work out for the species through time and others that may not. Immigration, or sharing of genes, between populations ensures the genetic continuity of the species while not swamping out local adaptations.

An excellent example of this in geological time can be seen in evolution of complexity in the first lower molar of the North American meadow vole, *Microtus pennsylvanicus*. Anthony Barnosky and Leo Carson Davis documented changes in a number of features of this molar through time. Barnosky's work showed that molar characters of meadow voles from different fossil localities changed at different rates through the late Pleistocene. Furthermore, Barnosky showed that different characters even in one population evolved at different rates, a further expectation from modern population genetics.

Davis documented an overall increase in complexity of first lower molars in meadow voles from the Great Plains during the past 250,000 years. Similar shifts in average dental complexity are not expressed in meadow voles from forested regions to the north and east. That is, the voles from the eastern United States, for example, demonstrate an average complexity more similar to the underived ancestral form. These considerations suggest that a species, particularly one with a widespread geographical distribution, will demonstrate swarms of phenotypes maintained by natural selection in the face of the opposite influences of mutation, immigration, and genetic drift.

Neutral Theory of Evolution

The field of molecular evolution is relatively new, beginning in earnest in the mid-1960s. One observation that launched a huge research program in America and elsewhere was that mutations resulting in changes in amino acid sequences in proteins appeared to occur in organisms at a constant rate. This work, begun by Emil Zuckerandl and Linus Pauling, suggested that the amount of genetic variation in organisms was therefore a function of time (**Fig. 2.13**). All other things being equal, organisms with more genetic variation than their closest relatives should be evolutionarily older. Thus, if one could measure the rate of mutations in organisms, one could then create a kind of *molecular clock* and work backward in time to establish times of divergence between related groups (e.g., between chimpanzees and humans or between the different modern races of humans). Radiometric dates from the fossil record would act as calibration points. Molecular data from living animals and morphological data from the fossil record of the same creatures are sometimes in accord and sometimes are not, but there are enough studies in general harmony that it has become clear that molecular biology can contribute greatly to our knowledge of organismal evolution, particularly in groups that lack a fossil record.

Within a few years of the beginning of the field of molecular evolution, a Japanese investigator named Motoo Kimura made a startling proposition; that most amino acid base substitutions in DNA (point mutations) that become fixed in populations through time are neutral with regard to adaptation (or "fitness") and that genetic drift, not natural selection, determines which mu-

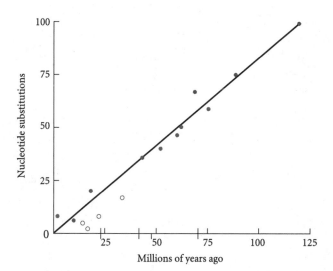

Figure 2.13 The theoretical relationship between mutation rate, as measured by nucleotide substitutions in seven proteins, and time in mammals. An assumption of such a constant mutation rate can lead to the construction of a molecular clock, allowing systematists to determine past times of divergence among organisms without use of fossils. However, mutation rates may not be constant, and phylogenies constructed from them must be viewed as only tentative until confirmed by fossil evidence. (From D.J. Futuyma, 1998. *Evolutionary Biology,* 3rd ed. Sunderland, MA: Sinauer Associates. p. 121; adapted from C.H Langley, and W.M. Fitch, 1974. An examination of the constancy of the rate of molecular evolution. *Journal of Molecular Evolution* 3:161–177.)

tations, and therefore which characteristics, are passed down through time. This model has become known as the "neutral theory" of molecular evolution. If true, the model would indicate that evolutionary changes in organismal characteristics would not be under the auspices of natural selection. Rather, the changes we observe in the fossil record would be the neutral outcome of a spontaneous mutation rate. Since this theory was proposed, many studies have identified selection in molecular systems, so the neutral theory is probably not a valid model for much of the observed variety seen in organisms. However, it serves as a useful model against which studies of natural selection should be compared. In claiming that natural selection is the driving mechanism behind observed change in a particular group, an investigator must first rule out the neutral model, or neutral hypothesis.

Macroevolutionary Patterns and Processes

When we step back and look at the long history of life on a grander scale, we no longer see the specific dynamics of individual animals interacting and reproducing from one generation to another. Instead, we see the cumulative results of those interactions and genealogical changes, and we see them imperfectly, predominantly because the fossil record is itself imperfect.

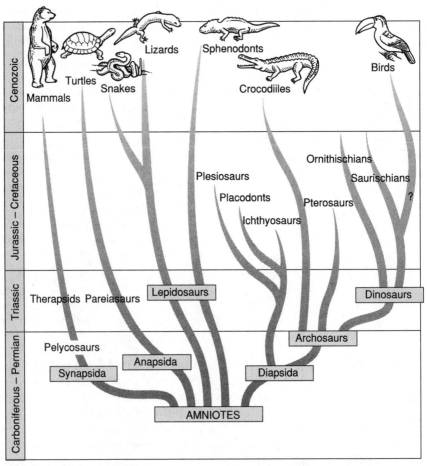

Figure 2.14 A possible phylogeny of amniotes, an example of a macroevolutionary pattern inferred from the fossil record.

Nevertheless, certain overall patterns emerge, and it is the job of evolutionary biologists to explain those patterns. For example, we can see that dinosaurs emerged from a group of early reptiles, flourished for hundreds of millions of years in the Mesozoic, and then the remaining taxa became extinct more or less simultaneously at the close of the Cretaceous Period (**Fig. 2.14**). A series of questions naturally arises for scientists: How and why did the first dinosaur evolve? What kinds of anatomical evolutionary trends characterize dinosaurs as opposed to other groups of reptiles? How many dinosaur species appeared and at what rate? As we know that many dinosaur species also became extinct during the Mesozoic, what was their rate of disappearance? Was it episodic, or continuous? What kinds of mechanisms can explain dinosaur extinction? Why didn't dinosaurs survive into the Cenozoic (i.e., why aren't there any dinosaurs alive today)? Are birds dinosaurs with feathers?

We also ask these questions about other animal and plant groups. Given a long geological time period and a limited fossil record, the patterns that represent the data associated with these questions are often presented as scenarios of ancestry and descent, sometimes with associated notes on morphological change. Or they may instead be graphical representations of origination and extinction, generally at the generic level or higher. From these scenarios we can see quite clearly that over geological time certain evolutionary tendencies prevail. We identify these tendencies toward directional size or structural change as *trends.* The largest horses with the fewest toes occur later in time than the smaller ones with more toes. The largest dinosaurs are also not the earliest. Many examples attest to the prevalence of size increase in clades through time. This general pattern has appeared so often that it has a name, *Cope's Rule,* after a famous American paleontologist who first identified this pattern in the fossil record. But what processes account for these long-term patterns? Can the process of natural selection at the individual level explain the patterns, or are other mechanisms at work? What is the influence of chance, versus directional selection, in the determination of trends? Is the change within a clade gradual or episodic? Are there any evolutionary reversals within a clade? That is, if a clade demonstrates a trend toward large size, do small species ever appear later in the clade's history? If so, how do we explain this phenomenon? Answers to some of these questions are presented in detail among the case histories later in this book and in the considerations below.

Punctuated Equilibrium and Phyletic Evolution

Twenty-five years is not a long time in the geological history of our species, but it is a long time for a novel scientific idea to remain in vogue. In 1972 Niles Eldredge of the American Museum of Natural History and Steven Jay Gould of Harvard University published a paper introducing the concept of "punctuated equilibrium." If evolution progresses in a slow, methodical, and gradual fashion, they asked, where are all the fossil intermediates between species? Perhaps they are difficult to find because they do not exist. Instead of a gradual evolutionary march from one species to another, maybe most evolutionary change in species occurs when they first originate and very little happens after that time. So *punctuation* refers to the origination of one species from another via a speciation event and the changes, anatomical or otherwise, that occur during or immediately subsequent to the speciation process. The *equilibrium* refers to the idea that after the initial shift of fea-

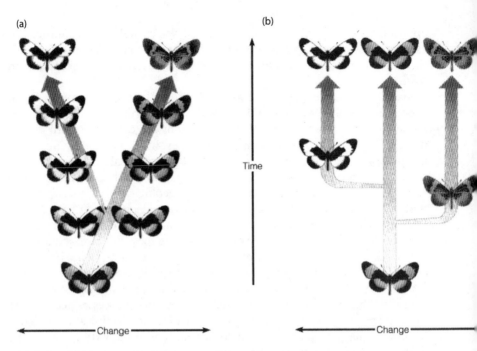

(a)   (b)

Time

Change   Change

Figure 2.15 The theoretical patterns of phyletic gradualism (a) and punctuated eqilibrium (b). Branching occurs in both cases, but in punctuated equilibrium most morphological change is restricted to a brief period during the origin of the new species, the speciation event. In phyletic evolution, change may occur (although not necessarily constantly) throughout a species' life span. (From *Biology: Concepts and Connections,* 3rd ed. by Neil A. Campbell, Lawrence G. Mitchell, and Jane B. Reece. Copyright 1994, 1997, 2000 by Benjamin Cummings, an imprint of Addison Wesley Longman, Inc. Reprinted by permission of Pearson Education, Inc.)

tures associated with speciation, characteristics of the daughter species change little, only meandering around average values for the rest of the species' lifespan. This net zero change is also referred to as *stasis*.

Microevolution can be summarized as natural selection working at the population level to shift organismal genotypes and phenotypes in directions that are most adaptive for living organisms. Darwin saw natural selection as a force that could change one species into another over time. This origination of a new species by evolutionary modification is often referred to as speciation by *anagenesis,* as opposed to speciation by *cladogenesis* (branching or splitting). Darwin did not deny cladogenesis; he recognized that life's full diversity cannot be explained without lineage splitting. Nevertheless, he viewed anagenesis as a valid and important mechanism, and many evolutionary biologists have assumed for a long time that species continue to change throughout their lifespans. This model of change, which is called *phyletic evolution,* results in a different visual pattern from that of punctuated equilibrium when viewed

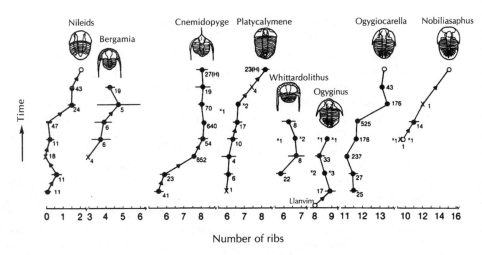

Figure 2.16 Evolution in the number of ribs in eight lineages of Welsh Ordovician trilobites over a three million year period, demonstrating a variety of phyletic patterns. Clearly, the changes in morphology among the species is not due to environmental changes that affected all in a similar fashion. (Adapted from P.R. Sheldon, 1987. Parallel gradualistic evolution of Ordovician trilobites. *Nature* 330: 561–563.)

on a graphical plot of morphological change over time. The resulting pattern of phyletic gradualism is a series of diagonal lines, branching at various points where speciation events occur (**Fig. 2.15a**). Each diagonal line (called a *lineage*) may also be broken into constituent species if, according to the investigator, significant morphological change has occurred to warrant this arbitrary fractioning. The fossil record of a group within which punctuated equilibrium dominates results in a graphical pattern that is more rectangular (**Fig. 2.15b**), and lineages are not broken into constituent species because, supposedly, no further evolution of significance is observed after the speciation event.

In the case of phyletic evolution, morphological trends can be explained by selection acting at the individual level, shifting populations gradually through time in a particular direction, for instance, toward larger size or greater complexity (**Fig. 2.16**). According to this model, little significant change occurs during a speciation event; speciation is seen as a method by which population fragmentation occurs and new evolutionary trajectories develop in daughter species because of a discontinuity from the ancestral gene pool. However, if most morphological change takes place within a very brief time associated with speciation, as postulated by the hypothesis of punctuated equilibrium, then long-term evolutionary trends must be the result of some mechanism other than selection acting at the individual level throughout the life span of a species. That is, species rarely continue to evolve in the sense that Darwin postulated. Eldredge and Gould and Steven Stanley, among other investigators, have developed models of selection acting among species in the same way that selection acts on individuals within species. Gould, in particular, has also championed the idea that trends may develop through random processes. Both concepts are expanded in sections below.

But what does the fossil record say about the prevalence of punctuated equilibrium versus phyletic evolution? Stanley and his colleagues demonstrated stasis for long periods of time in a number of marine bivalve species, and the punctuated record is seen in a few other marine invertebrate clades. The punctuated pattern has also been reported for a series of snail lineages from the Great Rift Valley of east Africa during the same time early humans were living there. However, the fossil record of terrestrial mammals often supports the phyletic model, but change within lineages is seen as episodic. That is, there are often long periods when change does not occur, followed by rapid evolution. These bursts of change are not associated with speciation events. In the case of the North American muskrat (considered in more detail in Chapter 9), the known history of which in North America spans a period of about four million years, the animal evolved from about the size of a small rat (ca. 100 grams) to its current size of around 0.8 kg. Some anatomical changes accompanied this size change. So there is support for both models. It remains to be seen which will turn out to be most common.

Selection Above the Species Level

If natural selection on individuals within species can set directional trends, is it possible that another form of selection could act on species within clades? That is, could a species be a kind of higher order individual and therefore be subjected to the same forces that act on an individual within a species? Michael Ghiselin has argued from the purely philosophical point of view that species are individuals; accepting this premise for the moment, what are the ramifications? Almost 20 years ago Steven Stanley proposed the term *species selection* for a process that acts on species rather than simply the individuals that compose the species. He came up with this idea in part because he was convinced that phyletic evolution was too slow to account for the morphological diversity of organisms on the planet. Since then this idea has been challenged and modified, but for the purposes of this treatment I am including any effect acting at the species level under Stanley's term. So how is it possible that an evolutionary influence could operate on a species without acting on the individuals that make it up? This is a difficult idea to conceptualize, but it can be done if we focus on the properties of a species considered as a collective entity.

Consider the cotton rats mentioned previously. They disperse in small breeding groups known as *demes*, and the tendency toward population fragmentation is fairly high as a result. But this tendency is not an individual property like size, shape, food choice, or physiological response. Another feature, population size, is also not an individual property, though of course it is determined by the collective reproductive rates of individuals. Both of these properties can have a profound effect on evolution and extinction within a clade. For instance, there will be more species in a clade that is prone to higher levels of population fragmentation. That is, its *speciation rate* will be elevated. In general and with all things being equal, an individual or group with a higher reproductive rate will replace individuals or groups with a lower rate; this is a well-known principle of population biology that has been verified many times in the laboratory. There is also a suggested link of extinction with population size. Species that have a tendency toward small populations may be more prone to extinction than those with larger populations.

The point that proponents of species selection make is that trends within clades may have less to do with natural selection among individuals for resources and regional adaptation than it does with species level properties. By this argument, the trend toward large size in cotton rats could come about solely because larger cotton rats have a higher speciation rate or a higher population size than small ones. Cotton rats are probably not very good candidates for species selection because there is considerable evidence that competition between individuals is important to their success; also, the trends toward greater speciation rate and population size should, at least theoretically, be lower in the species of larger size.

There is at this point almost no compelling data in support of species selection or selection at any higher level for that matter, but it is a theoretical model that is intellectually interesting. One of the primary problems limiting its popularity is the difficulty of recognizing its effects in a spotty fossil record.

The Drunkard's Walk: The Potential Influence of Random Processes on Evolutionary Trends

Stephen Jay Gould of Harvard University was one of the most outspoken critics of the so-called adaptationist program, which he claimed vastly overemphasizes the role of natural selection and adaptation in organismal evolution, particularly in the explanation of evolutionary trends. Instead, he suggested that many observed trends are the result of random processes. For many years biomathematician's such as David Raup and Fred Bookstein have been warning that some observed patterns in the fossil record, often attributed to directional selection, can also be caused by random processes. In his new book, *Full House: The Spread of Excellence from Plato to Darwin,* the late Stephen Jay Gould uses the fossil history of deep sea planktonic foraminifera (colloquially called "forams"), tiny protozoans with often elaborate inner or outer skeletons, to demonstrate this potential. If one examines the history of these forams over the past 65 million years, there is a distinct increase in average size. However, when the record is dissected and carefully analyzed, it can be determined that in any particular time period there is just as likely a chance that a new species will arise smaller than its ancestor as larger. If this is true, how can a trend toward large size develop? Gould uses the "Drunkard's walk" analogy in explanation. Suppose a drunk appears outside a bar. He is standing on a sidewalk 30 feet wide. There is a wall on his left and a gutter on his right. He begins staggering, and each stagger takes him either 5 feet toward the wall or 5 feet toward the gutter. His drunken bout is over if he reaches the gutter, at which point he falls over in a stupor. Now, if he begins staggering randomly in 5-foot increments, where will he end up? In the gutter every time, because the left wall acts as a "reflecting boundary," allowing no further movement in that direction. The only direction of continuous motion available is toward the gutter. Now, it may take, like flipping a coin, the equivalent of flipping six heads in a row (1/64 chance) to get there, but get there he eventually will. Similarly, Gould explains the evolution of large size in forams as the random evolution away from small size, not the positive selection toward large size. Forams get larger because that is the only overall direction in which their diversity can expand. As Gould and others have demonstrated, data from the foram fossil record show that small species are always common; the modal size (most common size, as opposed to the average size) of forams remains the same for 65 million years as larger species begin to appear.

The evolution of forams offers a likely scenario in which random events may actually explain a trend, but in other cases selection seems to operate. In 1986 I published a study on the fossil record of cotton rats over the past four million years. Cotton rats are small rodents, found mostly in southern grasslands of the United States, Mexico, Central America, and South America. As in forams, this clade also starts out small. And small species are occasionally found, even today, and did arise at random through the four million year study period. However, unlike in forams, the small species were more prone to extinction than the large ones. Today we have only one small species and many large species. Cotton rats are highly aggressive animals, and I believe the trend develops because larger species are more dominant in aggressive encounters and eventually replace the small species wherever they coexist. Cases of cotton rats attacking and killing pups of other rodent species have been documented.

So where does this leave us with regard to neutral or random evolution? Neutral characters certainly exist; they are, after all, the stuff from which Nature will select in the future. Random walks also may be more common than we suspect. They may dominate evolution in a clade after the appearance of a new species with a key innovation that effectively transfers the animal from one adaptive zone (or "ecological niche") to another. I have elsewhere labeled this occurrence as a "first-order speciation event" to separate it from the series of speciation events that follow, normally labeled (though mislabeled) "adaptive radiation." The diversity of species in a clade that results after a first-order speciation event is probably mostly random in direction and neutral with regard to adaptation. Apparently, one planktonic foram is just as functional as any other. This does not appear to be the case in cotton rats, but it may be true in many clades.

Origin of Species

Darwin recognized that life's full diversity cannot be explained without lineage splitting, known also as *branching,* or *cladogenesis.* For the purposes of this treatment a lineage is considered synonymous with a species. As mentioned above, species are born, they live, and they die (become extinct). A lineage is an arbitrary line drawn on a two-dimensional surface representing the genealogy of a species from its birth until its death. The line is composed of points, each representing a generation. Although it is true that a generational point also connects an ancestral with a daughter species, these generations are inherently different from ones that do not lead to lineage splitting because in the former breeding between successive generations ceases, whereas in the latter it does not.

Consider a species of rodent with a widespread geographical distribution during a cold period of time in the near past, approximately 600,000 years ago, known in North America as the Kansan glacial period. Let us say that this rodent species was found during this time in the valleys and the mountains of the Appalachian chain. During the warm interglacial period that followed, the glaciers retreated and small populations of this species became isolated on various mountain tops because the rodent's preferred habitat and thermal requirements were maintained. In a subsequent period that saw the continental glaciers once again encroach on the northern United States, these populations again came into contact with each other as the rodent's habitat spread down into the valleys. Only this time they did not recognize each other as the same species and would not interbreed. Branching, or speciation by cladogenesis, had occurred. This scenario represents an example of what has been termed the *allopatric model* of speciation (**Fig. 2.17**). The term allopatric means geographically separated. Another term, *sympatry,* means that the ranges of two populations overlap in space. The allopatric model of speciation describes the situation in which populations are geographically separated for some period of time and when they come into contact again (become sympatric), they no longer interbreed. They no longer share their DNA. This is the most commonly accepted model of speciation, although there are others. It is very likely, for instance, that many plant and some animal species, particularly insects, have speciated within the ranges of their ancestors, a process known as *sympatric speciation* (cases of this sort may be happening

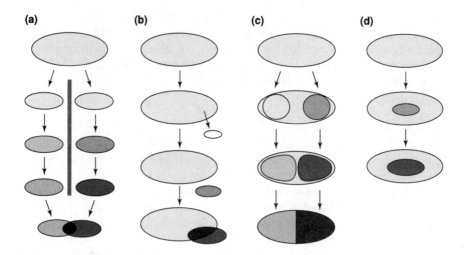

Figure 2.17 Illustrations of four models of speciation. (a) Allopatric speciation by vicariance—a barrier develops between two populations, and over time they diverge and do not interbreed when later sympatric. (b) Peripatric speciation (founder effect)—a small peripheral population becomes isolated from the parental population, as in dispersal to an island. (c) Parapatric speciation—large populations of a species diverge within the ancestral species' range by adaptation to different environments. (d) Sympatric speciation—one species originates from within the range of another, usually through differences in microhabitat choice. (Adapted and modified from D. J. Futuyma, 1998. *Evolutionary Biology,* 3rd ed. Sunderland, MA: Sinauer Associates. p. 483.)

now and are discussed in Chapter 12). However, this may be more a semantic than real distinction, as even in those cases where it is postulated there is a recognized physical isolating factor. It might be that a daughter insect prefers a new kind of plant to live on, or a new plant species prefers to root on a new type of soil. We can call this microallopatry, to distinguish it from the condition in which there is no overlap of geographical range between ancestor and descendant, but the principle is the same. There must be a barrier that stifles interbreeding of the potential descendant with the ancestral population for a period of time. It is during this period that the hereditary program of the descendant becomes shifted, or reordered, from that of its ancestor.

However, there is no *a priori* reason why populations that have been isolated cannot interbreed when they once again come into contact. They may indeed introgress, in which case speciation has not occurred. Perhaps some populations will interbreed and others will not. This is to be expected, because populations, especially within geographically widespread and polymorphic species with a patchy distribution, do not all evolve (change) at the same rate. This rate independence of populations is known as *mosaic evolution* and is predicted by principles of modern population genetics. As noted previously, Anthony Barnosky demonstrated this principle in fossil populations of the meadow vole over the past 30,000 years.

Recognition of speciation in the fossil record is not easy. Without the aid of a time machine we must rely on circumstantial evidence based on parts of animals that fossilize. In the case of mammals, statistical comparisons of skeletal and dental features become proxies for reproductive experiments. Some scientists call species in the fossil record "paleospecies" or "morphospecies" to distinguish them from "biological" species and to make the point that information about their biology is necessarily limited. The most pessimistic refuse to use paleontological data to determine evolutionary relationships among species, claiming that one can never determine an ancestor-descendant species relationship for sure based on fossil materials. The fossil record, particularly of the last few million years, is now becoming so dense that we can determine many ancestor–descendant pairings with a high degree of certainty. With the obvious limitations imposed by the passing of time and the presence only of skeletal remains, we can tentatively apply Mayr's species concept to the fossil record. It may be that some fossil samples represent mixtures of more than one generation, but there are modern statistical techniques, such as examining the range of variation expressed in the samples, to determine if mixing has occurred and to what extent. We can also use variation in modern samples of known species as a control to the extent of variation we see in the fossils.

In actuality, recognizing species in the fossil record is not as bad as some make it out to be. There is a simple rule that can help. If followed, it might also be a useful tool in unraveling the morass of taxonomic names among the many hominid species. All paleontologists should begin with a basic *null hypothesis* of relationships between fossil samples. A null hypothesis is also known as a *no difference* hypothesis. It is the standard hypothesis that one tests for in an experimental procedure. Given two fossil samples, one should begin with the idea that they are really the same, and only after rigorous statistical testing disproves this hypothesis should one consider that they may represent different species. Furthermore, if two distinct fossil samples, representing potential ancestor and descendant species, cannot be documented from the same locality or from contemporaneous but geographically separated localities, then the investigator should suspect that they are not different species.

Some scientists, myself included, suggest that speciation as a process should be limited to the condition in which a net increase of species occurs. This can be recognized only by contemporaneity of the parent and daughter species. Thus "anagenetic speciation" (speciation within a lineage) becomes an oxymoron. No matter how much change occurs within a lineage, because there has been no branching, speciation has not occurred. Evolution has occurred for certain, but no speciation. Speciation is not evolution; it is a form of population fragmentation that severs reproductive continuity between ancestral and descendant populations and generally initiates evolution along new trajectories in the descendant. Mayr was absolutely right: The salient point in the biology of species is whether or not their populations exchange genes. So how do we handle all those messy intermediate conditions—the rosencreis species, hybrid zones, and subspecies? As best we can. We do not need to panic when we see these phenomena; they are completely predictable by the principles of mosaic population expression seen in a long-term chronological framework, some examples of which are documented in later chapters of this book.

Species Extinctions

Extinction seems to be proportional to origination. That is, those species that have the fastest speciation, or origination, rates also have the highest extinction rates. This makes some sense or else the world could be populated only by species of mice or small insects. The British biostatistician J. B. S. Haldane is credited with having suggested that God must have an inordinate fondness for beetles. Beetles are among the most diverse animals on Earth, and it is an interesting coincidence that Charles Darwin, in his younger days at Cambridge University, also had a fondness for these creatures. But what happens to species that they should become extinct? There is a limited set of possibilities: They are displaced through competition, they become eliminated by climate and habitat modification, or they evolve into new species. If they are eliminated by climate, the change can be somewhat gradual or relatively instantaneous, as in a geological catastrophe of some kind.

Competitive Extinction

During the early middle part of the twentieth century, competition theory influenced a number of biologists to construct simple mathematical models of species interactions that would lead to precise outcomes, notably either coexistence or extinction of one of a pair of species set up in competition. These are known as the *Lotka-Volterra* models, and they were tested first in the laboratory in what are euphemistically referred to as "bottle experiments," with microbes such as *Paramecium*. From early on it became clear to the investigators that competition was a whole lot more complicated than had been surmised. Sometimes the competitors coexisted, and sometimes they did not, often depending on the nature of the "bottle" environment. It was not until the 1960s when a number of scientists began to carefully study species interactions under natural circumstances that competition was shown conclusively to occur. Joseph Connell's work with marine barnacles is a classic. He showed that both environmental influences and direct aggressive behavior between species maintained the distinct zones inhabited by two species. One species would actually pry the other off the rocks, but the more aggressive species was also less able to tolerate desiccation at low tide, and so the less aggressive species was able to survive in a zone that was drier for a longer period of time. There are many other examples of competitive encounters, some very obvious and others more subtle. Cotton rats of the genus *Sigmodon* will attack and kill pups of other species in their habitat. Lions will kill any other cats they catch in their territories.

The fauna of South America housed a fabulous array of marsupial mammals for much of the Cenozoic Era, the last 65 million years. The marsupials of South America were equally as diverse as those found today in Australia. However, about two million years ago a connection developed between North and South America, and many terrestrial placental mammals invaded South America. Almost all the marsupials became extinct, with the exception of a few oddballs and the ubiquitous "possum." It is generally believed that the placentals outcompeted them. Marsupials are known to have lower metabolic rates than placentals at comparable body size and also generally lower reproductive rates. Circumstantial evidence in support of this scenario is the survival of a diverse marsupial fauna only on the island continent of Australia, where placental mammals other than bats never arrived.

The fossil record of small terrestrial mammals from Kansas shows that at various times new waves of immigrant species entered the region. Studies we have been conducting at Murray State University indicate almost continual replacement of rodent species in Kansas throughout the past 5.0 million years. These replacements appear to be due to a combination of habitat restructuring and competition. Early species of voles, Family Arvicolidae, were replaced completely by around 0.67 million years in Kansas by more progressive species of the living genus *Microtus*. Species of this genus are known as the "*Drosophila*" of the paleontologist because they speciated so rapidly and demonstrate considerable morphological change within lineages. They are ideal subjects for evolutionary investigations, and many scientists worldwide have been studying them for many years. They reside today almost exclusively in fields, prairies, and forest edges. Relics of the earlier radiation of less derived voles survive today only in forested aquatic or marginal habitats.

Effects of Climate

Any observant gardener can testify to the effect that weather changes, either during a single year or over the long haul, have on plants. Some years are extremely dry, and certain species die or grow poorly. In other years there may be a bounteous crop. Likewise, on scales of hundreds, thousands, and millions of years, climate can exert its effect on organismal survival, including entire species.

By far the most likely culprit for species extinctions is climatic change. Species change through time as natural selection operates on them and as environments change, but many times the changes are just too drastic. As Steven

Stanley discussed in his excellent book, *Extinctions,* major climatic changes have been seen as the likely forces in many periods of extinction throughout the entire history of life on Earth. Some investigators, such as K. D. Bennett of Cambridge University, believe that most extinctions occur at particular periods of time that are tied to extraterrestrial cycles. This is because of some very straightforward relationships. The Earth gets its heat (or *insolation*) from the sun. But the amount of heat it receives is dependent on its axial tilt and the nature of its elliptical orbit. Both have changed within recorded time, and these changes have allowed some investigators to determine the extent and periods of these changes in the past. Equations predict that the changes will have a periodicity; that is, will be repeated at particular intervals. The intervals appear to occur every 400, 100, 41, 23, and 19 thousand years (Ky), mostly due to the gravitational influences of Venus and Jupiter. These orbital variations are often referred to as "Milankovitch cycles," after the Serbian engineer Milutin Milankovitch, who dedicated much of his life to their study. Bennett and others point to many examples of rhythmic sediment accumulation and changes in the oxygen isotope record to support the orbital forcing hypothesis.

The modern distribution of plants and animals is only a relatively recent development. During the past 2.5 million years (My), from about the time that ice rafting is first documented in the late Pliocene fossil record of the oceans, there have been many oscillations between warm and cool periods that generally follow the Milankovitch periodicities. The major cool periods, known as *glaciations,* have been followed by warmer periods, known as *interglacials.* During the glacial periods, sea level also dropped considerably. For example, during the last glacial period (the Wisconsinan of North America; the Würm of Europe) sea level dropped more than 90 meters (about 270 feet). Conversely, coastal regions were flooded during the interglacials. Each time a glacial period occurred, much of the Earth's precipitation was trapped as snow at the poles and on mountain tops, and great continental and alpine (mountain) glaciers migrated southward and to lower elevations. Much of today's topography in northern latitudes is due to the activity of these icy masses. Axiomatically, each time the glaciers moved through, for example, Canada, it wiped the slate clean of living things. So the entire biota of the Arctic is only a few thousand years old. Even at points south there were dramatic effects of these climatic modifications, and many species of plants and animals have determined their distributions only within the last 10,000 years, and sometimes less.

The marine bivalve record, especially that of the western Atlantic, demonstrates similar extinctions during the late Pliocene and Pleistocene, though they stabilized into modern species about 125,000 years ago, during the last great interglacial period. Elisabeth Vrba of Yale University also showed that antelope species in Africa suffered a severe decline in abundance around 2.5 million years ago (Ma) during a period of global cooling. She postulated that certain bursts of speciation and anatomical change in animal groups are correlated with major rapid changes in climate and calls this idea the "turnover pulse" hypothesis. It is currently under scrutiny, and not all animal groups that have been studied support the model.

"Extinction" by Evolution

Species do change through time, and by this process they can sometimes modify their morphology and size significantly and move from one ecological adaptive zone (or "niche") to another. The North American muskrat did just this (see Chapter 9), and similar levels of change have been documented in a number of animal groups. Many paleontologists choose to arbitrarily break these kinds of evolving lineages into constituent species based on the idea that the differences among successive chronological populations are greater than the extremes of variation seen among related living species. However, there are serious philosophical and methodological problems with this approach, because populations within a species existing at different intervals of time are not ontologically equivalent to populations distributed geographically within a single time period. I will not bore the reader with more on this topic, but this recognition will eventually have considerable influence on both the way paleontologists do their naming and the counting of numbers of extinct plants and animals. The main point for this discussion is that organisms do not go extinct through evolution, and one must be knowledgeable enough about the fossil record to identify and then omit "pseudo-extinction" from one's analysis.

Mass Extinctions

Whether or not Milankovitch cycles are responsible for all major biotic change, extinction seems to take place on two scales: within the periodicities of the Milankovitch cycling, namely on the order of thousands and hundreds of thousands of years, and then again on the order of millions of years in what are referred to as *"mass extinctions."* Mass extinctions are recognized by their global effects of extraordinary magnitude. For example, perhaps the greatest single mass extinction occurred at the end of the Permian Period, the boundary of the Paleozoic and Mesozoic Eras. Approximately half of all marine families and many terrestrial organisms became extinct. Although the exact cause is difficult to determine, a reduction in rate of tropical limestone accumulation and a dramatic decline mostly in organisms with known tropical affinities suggests climatic cooling as the culprit. Global climatic change, seen as a major catastrophe occurring over thousands or perhaps even millions of years, is identified for most of the Earth's other periods of mass extinction as well. However, this does not appear to be the immediate cause for the great extinction event that occurred at the end of the Cretaceous Period of the Mesozoic Era, about 65 Ma. That extinction seems to have been caused by an extraterrestrial horse of a different color, like a bomb from the sky.

The world has not really been quite the same since the publication in 1980 of an article by the University of California (Berkeley) research team headed by Nobel laureate Louis Alvarez in *Science* magazine documenting a likely extraterrestrial impact as the cause for the extinction of the dinosaurs. The Alvarez group began their work in Gubbio, Italy, primarily to determine the rate at which limestone sediments had been laid down in times past. Limestone accumulates as the result of the death of billions of tiny sea creatures with shells made up of calcium carbonate, or limestone. When sea levels drop, or the land rises, these ancient sediments dry out and become exposed on land. Near Gubbio, there exist a series of limestone layers that just happen to bracket the end of the Mesozoic Era and the beginning of the Cenozoic Era, at about 65 Ma. The specific boundary layer separating the Cretaceous Period sediments from those of the Tertiary above (or the "K/T" boundary) is represented in the outcrops as a thin band of clay about the thickness of a quarter. What the Alvarez team was looking for was some kind of "clock" mechanism that would allow them to determine how much time was represented by the thickness of limestone in front of them. They needed something with almost an intrinsic rate of decay, such as a radioisotope, but there were no isotopes in the sediments. So they decided to check an "extraterrestrial" clock, the constant rain of certain rare metals from space. In particular, they chose to examine the proportion of iridium in these sediments. They knew the rate at which iridium was accumulating today, and if the not unreasonable assumption was made that iridium had always been bombarding the Earth at about the same rate, then the amount of iridium in the sediments could, in theory, determine how long it took to accumulate the sediments. What they did not expect was the tremendous spike of iridium they found in the clay layer representing the K/T boundary. Iridium could not accumulate to that extent under standard conditions. One of the few forces that could concentrate iridium to such a level was an impact from an extraterrestrial object, an asteroid. To solidify their position, the Berkeley team needed to demonstrate that the iridium spike could be found in deposits elsewhere of the same age, and this they were able to do. There are other ways in which iridium can be concentrated, both biological and geological, but to this point nobody has been able to seriously dent the Alvarez team's data. An alternate hypothesis identifying volcanic activity as a likely culprit of the iridium spike was proposed by Charles Officer and Charles Drake of Dartmouth College, but as

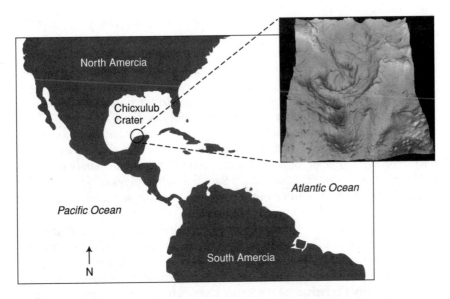

Figure 2.18 Seismic data showing form of the Cretaceous Chicxulub asteroid crater on the Yucatan peninsula. This impact occurred at the same time as the demise of the dinosaurs and many other organismal groups at the close of the Mesozoic era. (Photo inset courtesy of NASA.)

the years pass the evidence just keeps mounting in favor of the asteroid hypothesis.

Studies have now identified the impact crater for the event, the Chicxulub structure, located at the edge of the Yucatan Peninsula and extending somewhat into the Gulf of Mexico (**Fig. 2.18**). Geological evidence for the impact is unassailable. When an asteroid hits the Earth at tremendous speed, the energy of the impact causes a number of tell-tale physical changes in the geology at and near the impact site. First, of course, is the crater itself. Then the tremendous heat melts crustal rocks, sending up a spray of tiny glass-like spheroidal particles known as *microtektites*. A third piece of evidence is the presence of "shocked quartz grains," with a characteristic fractured appearance that can only be caused by the rapid application of extraordinary forces generated by something like an asteroid impact. Fourth is the presence of soot-like carbon particles from K/T boundary layers worldwide, interpreted as the remains of global wildfires initiated by the impact. Finally, there are geological sediments on Caribbean Islands representing tidal wave effects dated to exactly the time the impact occurred. Assuming for the moment that the impact occurred, can we be certain that it was responsible for the demise of many organisms at the end of the Mesozoic Era?

Was it specifically responsible for the extinction of the dinosaurs and the rise of modern mammals?

The environmental conditions generated by the blast of an asteroid perhaps 6 to 10 miles (10–16 km), wide would rival the most horrific nuclear winter scenarios suggested by war analysts. A huge dust cloud would blanket the Earth, reducing the earth's temperature to the point that most photosynthetic activity would be seriously affected. Evidence for biotic effects of the K/T impact is somewhat controversial, because it has been suggested that many groups of organisms were experiencing declines well before the impact. However, some of these studies may have to be redone in light of recent field studies of dinosaur fossils. Philip Signor and Jere Lipps of the University of California at Davis pointed out that the poor records of many groups near the K/T boundary might give the illusion of a more gradual, or sequential, extinction than actually occurred. Indeed, further detailed collecting of dinosaur materials has shown that most dinosaur taxa persisted right up to the time of the impact. In general, the K/T boundary saw the demise of the dinosaurs and a number of marine invertebrate groups, in particular the ammonoids, extinct relatives of the living chambered nautilus, and a group of tiny creatures called calcareous nannoplanton. Terrestrial plant communities were dramatically affected; at the K/T boundary in the western interior region of North America (Wyoming, Montana, Alberta) there was a switch from angiosperms and conifers to ferns, followed by a rapid return to arboreal pollen. This change is interpreted as a response to dramatic cooling and is consistent with an impact scenario.

Suggested Readings

Anderson, M. 1994. *Sexual Selection*. Princeton University Press, Princeton.

Avise, J. C. 1994. *Molecular Markers, Natural History and Evolution*. Chapman & Hall, New York.

Bell, G. 1997. *Selection: The Mechanism of Evolution*. Chapman & Hall, New York.

Boag, P. T. and P. R. Grant. 1981. Intense natural selection in a population of Darwin's finches (Geospizinae) in the Galápagos. *Science* 214:82–85.

Carroll, R. L. 1997. *Patterns and Processes of Vertebrate Evolution*. Cambridge University Press, New York.

Eldredge, N. and S. J. Gould. 1972. Punctuated equilibria: an alternative to phyletic gradualism; pp. 82–155. In T. J. M. Schopf (ed.), *Models in Paleobiology*. Freeman Cooper and Co., San Francisco.

Freeman, S. and J. C. Herron. 2001. *Evolutionary Analysis*. Prentice Hall, Upper Saddle River, New Jersey.

Futuyma, D. J. 1983. *Science on Trial: The Case for Evolution*. Pantheon, New York.

Futuyma, D. J. 1998. *Evolutionary Biology.* Sinauer, Sunderland, Massachusetts.

Gould, S. J. 1977. *Ever Since Darwin: Reflections in Natural History.* W. W. Norton, New York.

Gould, S. J. 1977. *Ontogeny and Phylogeny.* Harvard University Press, Cambridge.

Grant, V. 1981. *Plant Speciation.* Columbia University Press, New York.

Kimura, M. 1983. *The Neutral Theory of Molecular Evolution.* Cambridge University Press, New York.

Martin, R. A. and A. D. Barnosky (eds.). 1993. *Morphological Change in Quaternary Mammals of North America.* Cambridge University Press, New York.

Maynard Smith, J. 1998. *Evolutionary Genetics.* Oxford University Press, Oxford.

Mayr, E. 1963. *Animal Species and Evolution.* Harvard University Press, Cambridge, Massachusetts.

Milkman, R. (ed.). 1982. *Perspectives on Evolution.* Sinauer, Sunderland, Massachusetts.

Mitton, J. B. 1997. *Selection in Natural Populations.* Oxford University Press, Oxford.

Strickberger, M. 2000. *Evolution.* Jones and Bartlett, Boston, Massachusetts.

White, M. J. D. 1978. *Modes of Speciation.* W. H. Freeman, San Francisco.

3

Classification and Ecology: Making Sense of Nature

". . . we have come to realize that a classification can be more than a catalog; it can be a chronicle of the evolutionary history of life on Earth."

Classification and Phylogeny

A classification is a verbal or written compilation of things, the purpose of which is to create some order out of an otherwise bewildering array of objects on Earth. It is based on the simple premise that those things with similar characteristics are more likely to be related than those that do not share these features. A classification can be done with anything—automobiles, minerals, literature, paintings, or organisms. The earliest forms of informal classification undoubtedly arose among our hunter–gatherer ancestors, a matter of survival. Plants were "classified" based on defining features, including those that were poisonous and those that were not. Rocks were classified based on which ones fractured easily or made the best knives, spearheads, or hand axes. This lore was passed down from generation to generation in a predominantly oral tradition. Indeed, primitive societies today have their own names for the species in their environments. In this chapter we are concerned specifically with the classification of organisms and the search for underlying principles that bind classification to evolution. Individuals who build classifications are known as *systematists,* and the scientific study of classifications is known as *systematics.* Each formal name in a classification (genus, family, class, etc.) is also a *taxon,* and another name for systematics in the literature is *taxonomy.*

73

In 1735 Carolus Linnaeus, a Swedish physician turned naturalist, published a huge scientific account designed to be the first catalog of living things on Earth. He recognized that each kind of organism needed a distinct name that distinguished it from all others no matter where in the world that organism was discussed, so he invented the formal binomial, a combination of the genus and species. We are *Homo sapiens,* and by the end of this book you will hopefully have gained a greater appreciation of what exactly that means. Linnaeus is considered the father of modern classification, but he was undoubtedly influenced, either directly or indirectly, by the writings of Aristotle, the famous classical Greek philosopher. Aristotle did not have a modern perspective of evolution or even of variability in living organisms, but he was an astute observer of nature. His legacy unfortunately includes the "Scale of Nature," known also by many other labels (the "Ladder of Perfection," "Chain of Being," etc.), which purported to show how organisms were arranged from the simplest to the most complex, with humans invariably at the top (**Fig. 3.1**). According to Aristotle, these rungs on the ladder were fixed and unalterable, being as they were only imperfect representations of perfection, which occurred in an imaginary Platonic "World of Forms." Linnaeus, as well as most of his contemporaries, accepted this general condition for the world's living things. Linnaeus was caught up in description, not understanding or explanation. But we have come to realize that a classification can be more than a catalog; it can be a chronicle of the evolutionary history of life on Earth.

It is an interesting condition of life that there is diversity among organisms. Why is the Earth not blanketed by a single ancestral microbe? The answer to this question is the driving force of evolution; the Earth's environment is variable and ever changing. To survive over the long term, populations must change and enough diversity must exist to survive even the most dramatic environmental catastrophes that cause mass extinctions. Indeed, one could argue that speciation, which is not necessary for immediate survival, has evolved at least in part to allow living things to survive geological catastrophes. So, there is then a real connection between the diversity that we perceive and the processes that created that diversity. As we shall see, it is this important connection that has led us to the most widely accepted modern method of constructing classifications.

Phenetics and Phylogenetic Systematics

Living fishes and amphibians look quite different. Fishes have gills, scales, and fins, whereas most adult amphibians have smooth slimy skin, four legs, and lungs. Traditional classifications treated fishes and amphibians as separate classes of vertebrates: the Pisces and the Amphibia, equal in rank to other traditional vertebrate classes, the Reptilia, Mammalia, and Aves (birds). This

Figure 3.1 Aristotle's Scala Naturae, depicting the implied, fixed relationships of organisms then known to him and their relationships to each other and to humans, the latter which were believed to be closer to "perfection" than anything lower on the ladder. (Modified and adapted from E. Guyenot, 1941. *Les Sciences de la Vie: L'Idee d'Evolution.* Paris: Albin Michel.)

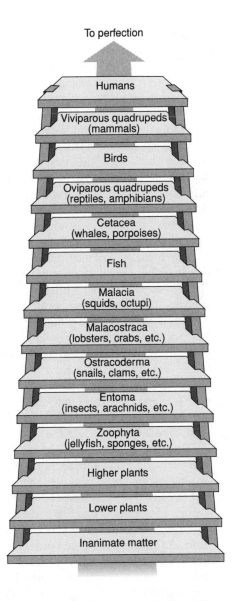

kind of classification is based on a *phenetic* approach to classification. The phenetic method classifies organisms on a statistical measure of distance between sets of characteristics used to compare them. Many, sometimes hundreds, of features are used. Information on evolutionary history is ignored. We now know that the actinistian (= crossopterygian) lungfishes, such as the "living fossil" *Latimeria chalumnae,* are more closely related to amphibians than they are to other "fishes." Another school of classification, known as *phylogenetic systematics,* or *cladistics,* would classify the lungfishes with the living Amphibia in a larger group known as the Sarcopterygii (see Chapter 10). It also turns out that the living amphibians are just one of a number of tetrapod (four-footed) animal groups that originated during the Devonian Period. Cladistics originated with the German biologist Willi Hennig, and according to his precepts, because lungfishes are linked to amphibians by special derived characters (e.g., limb homologues in the fins), they must be classified with them. Cladistics uses *phylogeny,* the evolutionary history of organisms, as the final arbiter of classification. Because all methods of classification must be based on comparing characters, and if organisms that share characters are generally more closely related than those that do not, then why do we have these two schools of classification and how is it possible that they can come to different conclusions about relationships among organisms?

Very often, the phenetic and cladistic approaches do result in the same classification. However, there are occasionally differences, as in the case noted above, because of their different philosophical basis. In phylogenetic systematics, the integral taxonomic unit above the species is called a *clade,* defined as a *monophyletic* group, or an ancestor and all of its descendants (**Fig. 3.2**). So, if we allow that the living lobe-fin fishes are related to living amphibians, whereas some other fish-like aquatic vertebrate was ancestral to all other bony "fishes," then according to the cladistic approach lungfishes should be classified with terrestrial vertebrates, not with other bony fishes. Birds, despite possessing feathers, may be more closely related to dinosaurs than they are to other reptiles. At least one cladistic classification lumps the birds and small bipedal theropods in the Coelurosauria, a dinosaur group (see Chapter 7). The primary point made by cladists is that there is no biological foundation to the phenetic approach, no way to assess which classification derived by phenetics is "real." On the other hand, there has been only one tree of life, and this gives the cladistic approach a goal: a single solution to organismal history. However, it must be remembered that any cladistic classification is only as good as the characters used and the statistical comparisons made.

In addition to their philosophical differences, phenetics and cladistics differ on how characters are assessed. Although some characters in phenetics may count more heavily than others ("weighted"), most often one character is treated pretty much as any other. The general idea is that with hundreds of characters, the effects of the unimportant ones will get diluted. Cladistics recognizes that only certain characters can be used in phylogeny reconstruction. Many features of organisms are shared with their ancestor and their relatives. These are useless for determining the direction of evolution and the relatedness among members of a large clade. For that, one needs to find characters that are unique and shared only by a particular group of related descendants.

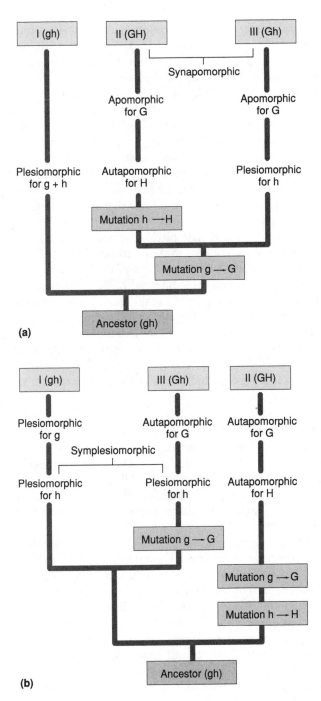

Figure 3.2 Two cladistic interpretations of the relationships of three organisms (I, II, III) based on various combinations of four characters (g, G, h, H). The gh condition is considered ancestral. Choice (a) is preferred over (b) because it includes one less mutation, or is more "parsimonious." Plesiomorphic = primitive, symplesiomorphic = shared primitive, apomorphic = derived, synapomorphic = shared derived, autapomorphic = unique derived.

These shared derived characters are called *synapomorphies*. In cladistics, only synapomorphies are allowed in the recognition of clades. Otherwise, one could accidentally include taxa in a clade that are unrelated. For instance, if we use the combined presence of gills, scales, and paired fins as the characters to classify vertebrate animals as the Class Pisces, lobe-fin fishes would be included. However, according to the cladistic approach, these features are generalized underived features (called *symplesiomorphies*) shared by all aquatic vertebrates at one time in the past and thus useless to determine who is really related to whom. To determine the relationships among these early vertebrates, one must take a closer look at features of the skull and limbs, and when that is done the relationship of lobe-fin fishes to amphibians becomes clear, and the remaining bony fishes, which have bony rays in their fins, are classified as the Actinopterygii.

In practice the identification of the correct phylogeny and thus classification is difficult, and it requires a computer to make comparisons between large data sets for many taxa. Programs that perform phylogenetic analysis such as PAUP and MacClade are available commercially. These programs all function on the premise that the biologically "real" classification is most likely to be the simplest, that is, the one that requires the least number of evolutionary changes. This is known as the principle of *parsimony*. So the programs all manipulate the data matrix, looking for the most parsimonious classification consistent with the fewest character changes. The underlying principle is that mutations leading eventually to new phenotypes are not reversible (also known as *Dollo's Law*).

Homology, Analogy, and Evolutionary Convergence

The theory of evolution rests on many mutually supportive pillars of evidence, one of which is the presence of *homologous* structures among related organisms. The assumption is that similar structures imply common ancestry, and wherever this has been tested in the fossil record it has been shown to be true. So the wings of an eagle and a sparrow are similar because they evolved from a common ancestor, whereas the wings of a bird and a butterfly are different because they do not share a recent common ancestor. Bird wings are also said to be *analogous* structures, because in addition to sharing a common structural ancestry, they also perform the same function in descendants. But they do not need to be homologous. For example, many studies from both the fossil record and developmental biology have shown that human limbs are homologous to the limbs of other vertebrate animals (**Fig. 3.3**). They may not be analogous anymore, because the limbs of other vertebrates are adapted to many kinds of environments and activities, but a basic structural plan is shared by all vertebrates because they share a common ancestry. If evolution and genealogical relationship were not the cause for these simi-

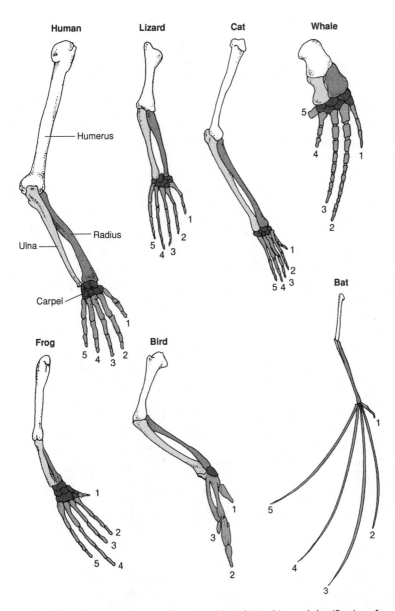

Figure 3.3 Systematics, the scientific study of the relationships and classification of organisms, is based on the assumption that common structural features indicate common ancestry. Although the forelimbs of the various animals illustrated here differ from one another, all share a common underlying bony pattern. These forelimbs are, therefore, considered **homologous,** and we can conclude from this that the animals with these limbs share a common Paleozoic tetrapod ancestor that was probably pentadactyl (with five fingers).

larities, there would be no reason for their consistency among groups. This is seen with spectacular clarity in the case history chapters; we are sometimes fortunate enough to have, in modern animals, remnants called *vestiges* of ancient ancestral structures that show the evolution of homologies in living species. As it turns out, the feet of ancient artiodactyl mammals are homologous to the flippers of whales. In part, this is why cladistics is such a powerful approach to classification; at its best phylogenetic systematics uses only homologous characters to infer relationship. But the flip side of homology, analogy, can occasionally be confusing and lead to erroneous classifications.

One of the great principles in ecological theory is that unrelated organisms may come to resemble one another if, in the course of evolutionary time, they are faced with solving the same ecological problems. That is why evergreen trees in the Rockies and the Alps look the same but are different species; why porpoises, sharks, and ichthyosaurs have streamlined body form and teeth of similar size and shape to catch fish; why butterflies, bats, and birds have wings; why some bats and hummingbirds have long snouts and tongues to gather nectar. The similarity of body form between a porpoise and a shark is an analogy, developed for moving swiftly through water. This is also an example of *convergent evolution,* in cladistic parlance called *homoplasy.* Phenetic classifications recognize that homoplasies can occur but have no consistent way to deal with them. That is why a purely phenetic classification will ally a crossopterygian lungfish with other fish rather than with amphibians every time. One of the benefits of the cladistic approach is that it better recognizes convergence than does phenetic methodology and thus limits the arbitrary grouping of organisms according to the whims of the investigator. After all, if there is no underlying biological rationale for a classification method, then the classification merely depends on the characters chosen by the authority. As mentioned above, one of the undeniable benefits of the cladistic approach is that there is only one "true" history of life on Earth. Cladists may modify their classifications as new information becomes available, but it is always with the intent of more closely approximating this biological reality.

Ecological Context

Sometimes ecology is portrayed as something separate from evolution, and at most universities the subjects are taught as if they share little in common. In part, this is because practicing ecologists and evolutionary biologists have classically been taught differently, and there is little overlap of subject matter in scientific journals and at national meetings. In fact, ecology is as much the study of natural selection on organismal phenotypes as it is a discipline that examines the interactions of organisms and their abiotic environment. In Chapter 2 we saw that both climate and competition may act as agents of selection. So when we examine the population biology, mating habits, community structure, temperature tolerance, dispersal mechanisms, habitat selection, and physiological responses of living organisms, in reality we are looking at microevolution in action, because these characteristics and responses of organisms are a reflection of the hereditary programs that lead them through the gauntlet of environmental challenges that fine tune them to their surroundings.

Many years ago Arnold Shotwell, a paleontologist, tried to make some sense out of the animal remains he found in ancient deposits. He made the working assumption that the animals that were represented by the most elements were those that lived nearest the fossil site, whereas those that were rarest were found, perhaps in a different ecological situation, farther from the site of deposition. He put the most common into what he called the *proximal* community and the less common into the *distal* community. For the time in which he worked, the 1950s, it was a valiant attempt to infer something about natural circumstances when the animals lived. Shotwell had the additional problem of working in deposits that were old enough such that no modern species were represented. Consequently, all of his information about habitat had to come from circumstantial evidence. We now know a lot more about the ecology of modern organisms and the way in which fossils accumulate, and Shotwell's work seems to us at this point somewhat naive. In fact, the situation is much worse than Shotwell imagined. There are many agents that collectively create a fossil assemblage, and it is not easy to tease them all apart to determine past environmental conditions and habitat structure. To show how paleontologists go about this process, we need information from both *taphonomy*, a scientific discipline concerned with the forces that are involved in creating a fossil assemblage, and ecology.

Natural History of Fossils

Fossils are not everywhere. Budding paleontologists, enthusiastic to go out and find a complete skeleton of *Tyrannosaurus rex* after seeing the movie Jurassic Park, need to first familiarize themselves with the kinds of rocks and the areas to look that will have the greatest probability for success. (And these days they also need to become familiar with the laws and etiquette of collecting. For example, one needs a special permit to collect on public lands and permission of the owners to collect on private land.) In general, fossils are found in *sedimentary* deposits. These are mostly fine-grained rock units that were laid down by either water or wind, as opposed to *igneous* and *metamorphic* rocks. Igneous rocks include such things as hardened lava flows, anything that was created under the extreme heat conditions of a volcano. Granite and obsidian are igneous rocks. Metamorphic rocks are those that have changed over millions of years, through the agents of heat and pressure. A well-known example is the diamond, formed from carbon.

Before we go any further we need to answer a simple question: What is a fossil? Are all remains from dead animals fossils? The answer is both yes and no. Obviously, fossils are remains of dead organisms. But geologists and paleontologists recognize that organic remains go through a process of change after a while. If left undisturbed, the organic material in, for example, a bone will be replaced by minerals. This process is called *fossilization* (**Fig. 3.4**). Nevertheless, the process is very uneven. Some bones thousands of years old, even of extinct animals, show little replacement, whereas others may be solid rock, with no organic material remaining. Mammoths and bison have been recovered from Alaska and Siberia, almost as if they had been "flash frozen." Are these fossils? It depends on how you are using the word. To some scientists, the answer is no. No replacement has taken place. Ergo, the mammoth is just dead, not "fossilized." On the other hand, it is extinct and thousands of years old. Colloquially it can be referred to as a fossil.

Fossils accumulate only under very special circumstances. The manner by which they accumulate and what happens to them afterward is the central focus of taphonomy. Let us consider a terrestrial vertebrate animal with bones. When most animals die, their remains are lost forever. Bone is, after all, organic matter and can degrade just like other parts of the body. It just takes longer. Exposure to the wind and rain, desiccation, and freezing can all break down bone. Scavengers carry off bones, and many mammals chew bones to keep their teeth growing properly and to obtain calcium. Bones can be broken and pulverized by trampling. So, for there to be a reasonable chance for preservation, potential fossils must be buried rapidly. This can occur under a variety of circumstances, especially near water. If animals die in or near water, there is a good chance their remains would be covered by mud or sand fairly quickly. And the anaerobic conditions that sometimes exist in these situations would also favor preservation. Conversely, very dry conditions are also good for fossils, because the agents of decay, bacteria and fungi, are minimally represented. Mummies hundreds or even thousands of years old have been found in dry caverns around the world. If animals are trapped by a sandstorm, burial will be rapid and complete.

Caves and sinkholes can accumulate tremendous numbers of fossils. Many animals live in caves throughout their lives. Cave communities are unique, including some very interesting and specialized creatures among the mammals, amphibians, crustaceans, insects, and chelicerates (spiders and scorpions). A number of animal groups have evolved special cave species that are albino and have traded eyesight for enhanced sensory abilities of other types, such as touch. The cavernicolous bats, of course, navigate by echolocation, the production of high frequency sound, or sonar. But caves are also inhab-

Figure 3.4 Fossils often accumulate and are preserved in aquatic habitats where they are protected from scavenging to some extent. Lower concentrations of oxygen and rapid burial in silt or sand also help. Fossils may remain in these sediments for millions of years until a new period of erosion or a combination of continental upheaval and erosion reveals them at or near the surface. (Modified from K. V. Kardong, 1998. *Vertebrates,* 2nd ed. Boston: WCB/McGraw-Hill.)

ited by animals that spend much of their time outside, and this can be the key to a tremendously rich fossil fauna. Many carnivorous mammals are known to frequent caves, such as coyotes, weasels, raccoons, mountain lions, skunks, bears, and foxes. These animals hunt mostly outside the cave but will drag their kills inside the cave. The remains of their kills are stockpiled in the cave and further bones and teeth are often excreted in their feces, in or near their dens. These dens may be occupied for thousands of years, often by quite different carnivores. After all, a good place for a fox is also a good place for a coyote. Given fairly dry conditions and a source of dirt to cover the material, a deep and rich fossil deposit can accumulate in cave sediments, especially in small isolated rooms. The "dirt" may be blown in or brought in by water, enter through cracks in the ceiling, or accumulate by erosion of the cave itself. Bat guano can also contribute to cave sediments.

Fossils accumulate by a variety of mechanisms, and their characteristics and position may change even after deposition. These days paleontological excavations are often conducted with at least the care and precision of an archeological dig. Sometimes the excavation is even more painstaking, as paleontologists are concerned as much with the smallest organisms as with the largest. So every inch of sediment is often sifted through the finest screens, on the order of cheesecloth, so as not to lose even the tiniest of elements. The smallest teeth in pocket mice of the genus *Perognathus,* for example, are less than 0.6 mm in diameter. The position and orientation of every element of every species is recorded and later reconstructed by computer in a three-dimensional diagram. In this way, scientists can look for patterns of deposition that may provide clues to the origin of the fossils. For instance, if laid down in moving water, bones may demonstrate a directional orientation consistent with current flow. If the current was swift, bones and teeth often show rounded and polished edges where they have been abraded while moving over other objects, such as stones.

The nature of bone fracture patterns can provide evidence of pre- versus postdepositional breakage. Sometimes bones are broken simply by the pressure of sediments and rocks on them over time. In other cases, however, breakage may have occurred before burial, perhaps due to carnivore activity or to human butchering. To differentiate between the agents in times past, it is often helpful to know what happens to modern carcasses when these forces act on them now. Anna K. Behrensmeyer's work with African mammals has been very helpful. She has shown, among many other things, that the bones of large mammals may be broken and gain many marks through processes that occur from the time an animal dies to when its bones are finally covered by sediments. With large herding animals, such as wildebeests, bones may be fractured and scratched by trampling of other herd members after the bones have been exposed and dissociated after decomposition. These scratches can mimic those caused by human butchering, so it is critical to carefully analyze the details of the scratches to determine which agent was responsible. Sometimes a scanning electron microscope is necessary to determine the difference.

How far do fossils move? This may seem like a silly question, but all professional paleontologists must be aware that the organisms they dig up in an excavation may not have been living together in the excavation area. Rich fossil localities are extraordinary circumstances and can be thought of as fossil "traps," or "sinks," that accumulate fossils for some unique reason. Consider a cave situation. Remains of organisms here represent two distinct sampling regions: the cave environment and the outside environment. Some of the fossils will undoubtedly originate from within the cave, but many others are brought in by carnivorous animals, including mammals and owls. The aggregation of fossils dug up thousands of years later is referred to as a "death assemblage," or *thanatocoenose* in arcane scientific parlance. It is quite different from the living assemblage, or *biocoenose,* both inside and outside the cave. In fact, the carnivores likely will forage through a variety of different habitats outside the cave. Canids and felids with large home ranges may traverse through two or three kinds of forests as well as prairies in search of food. One of the benefits of working with Pleistocene, or Ice Age, fossil beds is that the various communities can sometimes be identified because many of the prey species of the carnivores, mostly rodents, are living today and are restricted to specific habitat types. Thus, if a large proportion of the remains are from, for example, the North American meadow vole, *Microtus pennsylvanicus,* we can be certain that field or prairie ecosystems existed nearby the deposit.

As we go back farther in time and the vertebrate species become entirely extinct, our basis of inference decreases dramatically. We must then rely on other methods. One of the best is through analysis of fossil pollen, known as *palynology.* Modern species of plants have specific pollen types that often fossilize, and these plants have much longer fossil histories than animals. Even if the plant species are extinct, the proportion of wind-blown pollen, indicating grasslands, to pollen derived from trees can provide an important clue to the general habitats available. Elaborate pollen diagrams have been reconstructed from cores in ancient ponds and bogs, and they can tell us precisely how plant communities changed through a particular study period. Why don't we always use fossil pollen? Because in some sediments the pollen is either destroyed, leached out, or simply not present. This is regrettably the situation in most cave sediments.

As noted earlier, one of the best kinds of areas in which to hunt fossils is in old stream, lake, and river beds that have become exposed in arid regions, such as deserts. These are the kinds of ancient sediments in which the famous hominids from Olduvai Gorge, Kenya and Hadar in Ethiopia accumulated. However, one needs to be aware of the taphonomic processes at work here.

Aquatic systems are dynamic. Natural lakes, streams, and rivers have seasonal cycles that can dramatically alter fossil accumulations, sometimes leading to gross errors of interpretation. For instance, an older fossil bed can be *reworked* into younger sediments by the action of running water. This results in a mixed assemblage of older and younger materials. More than one paleontologist has been fooled into thinking a particular fossil aggregation was uniform when in fact it was not. And, of course, the carcasses of animals may be carried some distances from their place of death if the current was swift, as it often is during the spring floods, when many animals that live in a flood plain may die. Rich fossil beds may be found in ancient river oxbows, where the current would have trapped anything that came along. Fortunately, movement is rarely over very long distances, and the species in a fossil locality often really do reflect the kinds of environments nearby the site of deposition. Taphonomy has not negated paleoecological investigations; quite the contrary. Modern taphonomic practice actually allows us to learn a great deal more about the biology of our fossils when they were living, but we must be extremely careful in our investigations.

Ecological Inference

We started out in this chapter with the simplistic concepts of "proximal" versus "distal" communities. What is wrong with this idea? Well, to begin with, nature is not constructed this way. There is no reason to assume an even distribution of habitats around a fossil deposit. Rather, habitats have a patchy, or *mosaic,* distribution, with little predictability in geographical proximity to any one point. Also, thanks to the empirical and theoretical groundbreaking work of Robert MacArthur, G. Evelyn Hutchinson, and others, we have a much more refined appreciation of the population structure and distribution of organisms. In general, individuals within species are normally distributed in a community. This is the same statistical concept we saw in an earlier chapter, which simply means that the rarest species have very few individuals, the most common species have many individuals, and there are a few species that are extraordinarily common. So one cannot determine the habitat structure and distance from a fossil deposit by the *proportion* of individuals represented by fossil elements from the different species exhumed from the deposit, especially if all the species are extinct. It is difficult enough to do with a locality that is only 25,000 years old, in which most of the small mammals, for example, may be living species with distinct and well-known habitat preferences.

Consider the situation in which you have recovered the remains of two shrews, *Blarina brevicauda* (the short-tailed shrew) and *Cryptotis parva* (the least shrew), from a cave deposit. Suppose there are 100 specimens of the least shrew to 10 of the short-tailed shrew. Least shrews are most commonly found in open grasslands, whereas short-tailed shrews are generally forest dwellers. Can one rely on this proportion to determine a greater proportion of open rather than forested habitat nearby the locality? Based on this one species pair, such a conclusion would have to be very tentative. The shrew remains are almost certainly the result of carnivore activity. But which carnivores? Some carnivore species preferentially feed in grassland habitats where it might be easier to find prey. This is where, for example, most voles of the genus *Microtus* are found, and these rodents are often the dominant food taken by small and medium-sized carnivores. Thus, even if forests are near the cave, there may be an over-representation of grassland prey species because the predators preferentially fed in this habitat. So is interpretation impossible? No, but it is difficult. Many sources of information and techniques must be called into play to identify past habitats and climatic regimes. Some of these methods are described below.

Area of Sympatry and Indicator Species

Studies of modern animals show that it is the small ones that provide the most important information about habitat, because they are more restricted, or *stenotopic,* in their habitat choice. Because large animals need so much more energy to sustain them, they generally cross over into a variety of habitats. Their habitat selection is said to be wide, or *eurytopic.* So we usually look to the small animals, such as the shrews and rodents, frogs, salamanders and toads, snails, and insects, for help. We study the modern distributions of these species, and certain combinations generally imply certain habitats. For fossil accumulations within the last 500,000 years, where many of the species represented in fossil deposits are modern, we can make a fairly reasonable assessment of ecological conditions during deposition. One method that is useful in this regard is called the *area of sympatry* method with small mammals (though it can be done with any animal group or combination of groups). Sometimes, species are recovered in fossil deposits that are far from where they live today. These are called *disjunct* or *indicator* species, and they often send a powerful climatic signal. During the late Pleistocene, for instance, many northern disjuncts dispersed southward as northern habitats were extended southward at the front of the great late Pleistocene continental glaciations that covered most of Canada and the northern United States. If we examine the geographical distributions of all the small mammal species from a late Pleistocene fossil deposit, we can make the reasonable assumption that the area where all overlap today was likely representative of the overall climatic and habitat conditions during the time of deposition. Occasionally, confirmation may come from other sources of information, such as fossil pollen.

Although one must always be suspect of information from single species, some species send a stronger signal than others. The combination of Arctic shrew, heather vole, and caribou in a late Pleistocene cave deposit from Tennessee would argue forcefully for boreal coniferous forest, even in the absence of fossil pollen. Yet, we must be careful with these tentative conclusions, because we can never be certain what forces are responsible for the distribution of animals as we see them today. Perhaps humans overhunted caribou throughout the southern part of their range. Therefore, we must be especially cautious when inferring anything specific about past climate from these associations. We can say that the above combination of Tennessee fossils might imply summers that were on the order of 10° to 15°F cooler than today, but the evidence is certainly not conclusive. Better would be a more direct tie to temperature, and that may be forthcoming.

Oxygen Isotope Ratios

Fairly accurate estimates of past oceanic temperatures come to us from information imbedded in the shells of microscopic marine organisms, called Foraminifera. These creatures, members of the protozoa, have internal shells made of calcium carbonate, the same stuff that makes up limestone. Indeed, it is the shells of these animals and others that make up the limestones of the world. The chemical structure of calcium carbonate is $CaCO_3$. Water and the atmosphere contain oxygen in two stable forms, O^{16} and O^{18}. Because O^{18} is slightly heavier than O^{16}, more of it is taken into marine shells during colder conditions (warmer water and the lighter form of oxygen have a higher evaporative rate). So the O^{18}/O^{16} ratio is a very sensitive indicator of both water and air conditions (**Fig. 3.5**). We have used this ratio from shallow and deep sea sediment cores to construct a history of climate over the past four million years. From this we can see that climate has changed continuously, albeit episodically, during this period. If we could develop a method to examine stable oxygen ratios in terrestrial organisms, we could then have a direct reading of climatic conditions in a continental fossil deposit independent of the small mammal record. David Fox, of the University of Minnesota, will soon be investigating the use of oxygen isotope ratios in fossil rodent teeth from terrestrial deposits on the Great Plains of North America.

Pollen Profiles

H. J. B. Birks and H. Birks, in their excellent text *Quaternary Palaeoecology*, explain why pollen analysis has been one of the principal techniques used to reconstruct past environments, particularly those of the Quaternary, or the "Ice Ages," during the past 1.8 million years. Their reasons include the following:

1. Pollen grains are usually the most abundant fossils preserved.

2. Pollen grains are often resistant to decay.

3. The identification of pollen grains has been well established.

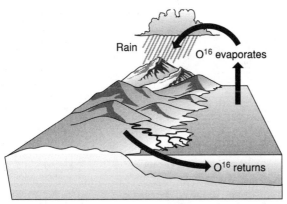

Rain

O^{16} evaporates

O^{16} returns

Interglacial: O^{18}/O^{16} ratio unchanged in ocean

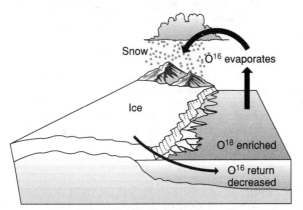

Snow

O^{16} evaporates

Ice

O^{18} enriched

O^{16} return decreased

Glacial: O^{18}/O^{16} ratio increased in ocean

Figure 3.5 The ratio of oxygen isotopes in fossil foraminifera taken from sediment cores at the bottom of the oceans can tell us about past climates. Under cold (glacial) conditions, more O^{16} than O^{18} gets tied up in ice after it evaporates from the surface of the oceans, so the oceans become relatively O^{18} enriched, simply because this isotope is heavier. Under warmer conditions, the O^{16} can return to the oceans. Because the proportion of oxygen isotopes in modern and fossil foraminiferan shells reflects the proportion of the oxygen isotopes in the water when they were alive, we can determine the temperature of the water (and, thus, the overall climate) during times past. (Modified and adapted from T. Van Andel, 1994. *New Views of an Old Planet: A History of Global Change*, 2nd ed. Cambridge, UK: Cambridge University Press.)

4. Only small amounts of sediment are needed because pollen grains are so small (5 to 100 μm).
5. Pollen grains originated from plants that grew together near the area of deposition.

Despite these positive aspects, there are some factors that must be considered carefully. All pollen does not disperse at the same rate and is not all produced in the same amount. For example, wind-pollinated species such as the pines generally produce much higher quantities of pollen than plant species pollinated by insects. An additional problem is that wind may carry pollen tremendous distances, so that some pollen from a fossil locality may not be from the local area around the deposit. Nevertheless, when many pollen profiles are compared over time, significant changes can be seen in absolute and relative

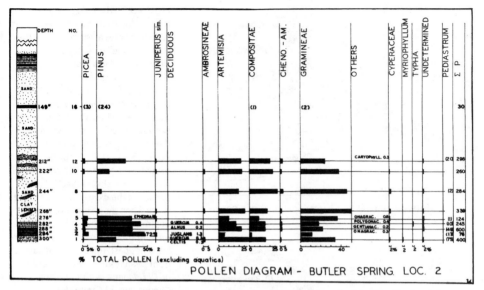

Figure 3.6 Pollen diagram for the Butler Spring late Pleistocene fossil locality from Meade County, Kansas. Lower levels show a dominance of pine and deciduous trees, changing to a drier period without arboreal pollen dominated by grasses, followed by a period when pine returns. The upper levels of the fossil deposit were mostly sandy and devoid of pollen. (From R.O. Kapp, 1965. Illionian and Sangamon vegetation in southwestern Kansas and adjacent Oklahoma. *Contributions from the Museum of Paleontology, University of Michigan.* 19:203.)

proportions that allow for a fairly clear identification of plant community structure and climatic conditions during the time of deposition (**Fig. 3.6**).

Cenograms and Body Size Estimation

Another more esoteric method to identify past habitats, introduced by J. Valverde and developed further by Serge Legendre in Montpellier, France, is called cenogram analysis. A cenogram is a graph of body mass (or weight) on the y axis plotted against the rank order of mass on the x axis for mammal species in a modern or fossil community (**Fig. 3.7**). It seems that the distribution of body sizes among living mammals can indicate the kind of ecosystem within which the animals are found. Cenograms for closed tropical forests include many species with virtually all mass classes represented. However, as one moves from a tropical rainforest to more open and arid environments, a distinct gap in the medium-sized species begins to become apparent, and the cenogram becomes steeper and shorter as fewer and fewer large species are represented. If this method proves reliable after study of many modern ecosystems, then it has considerable applicability to fossil assemblages, assuming one can accurately estimate body mass in extinct species.

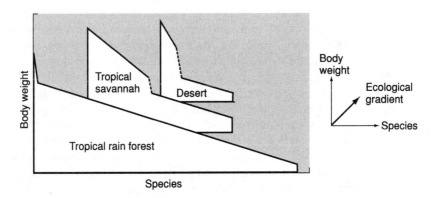

Figure 3.7 Cenograms, graphs that order mammal species in ecosystems by average body mass (weight). Average body mass is plotted on the *y* axis, and the species are then plotted on the *x* axis, from smallest to largest, by their average mass. In some ecosystems, such as tropical rain forests, there is almost a smooth continuum of species represented at all body sizes. However, as one examines more arid (drier) communities, such as grasslands and deserts, the number of body mass categories diminishes and there is an absence of species in certain size categories (dashed line). Methods are now available that allow scientists to accurately estimate body mass from dental and skeletal parts, and the construction of cenograms from these estimates can provide important circumstantial evidence for past habitats.

The techniques involved in estimation of body mass and correlated variables are examined in a book on the subject edited by John Damuth and Bruce McFadden titled *Body Size in Mammalian Paleobiology.* In this volume my colleagues and I demonstrate that accurate estimates of body mass in extinct mammals can be generated from equations based on the tight relationship between body mass and some dimension (e.g., a tooth size or limb measurement) in modern mammals. In addition to application to the cenogram technique, body size estimation allows examination in extinct species of other biological parameters correlated with mass in living species, such as metabolic rate, home range size, and population size (see Chapter 9).

Dentition and Diet: Where Form Sometimes Meets Function

The old adage "you are what you eat" can be rephrased into "you are what your teeth eat." Despite all the rhetoric about neutral characters and random processes in evolution, in most cases an animal's foraging style will be reflected by its mouthparts. This would not be possible unless there was active selection to refine this energy processing system. This can be seen clearly in the teeth of living animals, especially mammals. The teeth of carnivorous mammals generally have sharp cusps, called a *sectorial* dentition. Examples are weasels, cats, wolves, bats, and shrews. Even among the rodents, where there are a few rare species that exclusively eat meat, their cusps also tend to be sharper and more pointed than other rodents that eat less or no meat. Good examples in North America are species of the genus *Onychomys*, the grasshopper mice, which eat insects. Mice of the genus *Peromyscus* also include some insects in their diet, but their teeth are not as sectorial as those of *Onychomys*. Where bats have evolved to drink nectar, like hummingbirds, they have reduced the dentition enormously and the snout has become extremely elongate. Some carnivorous mammals have developed stout jaws and teeth to crush bones, such as hyaenas. Cats are the ultimate carnivores, and the upper and lower teeth are reduced at the expense of what is referred to as the "carnassial pair," the upper fourth premolar and the first lower molar. These two teeth function much as a pair of scissors, and little processing is done by other teeth. The skull and jaws are also shortened.

Grazing and browsing mammals, such as voles of the genus *Microtus*, deer, horses, gazelles, cows, elk, and so forth, have teeth designed to pulverize and shred plant materials. The premolars and molars of exclusive grazers generally flatten into two-dimensional planar surfaces. Primates, notorious herbivores, also have low rounded cusps on their molars. Most primates are fruit eaters and folivores (leaf eaters), though chimpanzees include a modest amount of meat in their diet, mostly colobus monkeys and an occasional gazelle. Chimpanzees and hominids in general are just about the only type of mammal for which one cannot clearly infer diet from dentition. This is very interesting and may express an important aspect of human evolution. We are not designed to eat meat, yet we do, in some cases almost exclusively. This occasionally results in appendicitis, because the short vermiform appendix associated with the beginning of the large intestine was not designed to store toxic byproducts of meat digestion. It seems likely that this switch in diet, almost surely of positive selective value because of the high energy content of fat and protein during early development, is an example of cultural rather than physical evolution and is associated with the more developed cognitive skills of humans. In a sense, the enigmatic case of human carnivory is the exception that proves the rule. In few other mammalian groups does one see this kind of contradiction. In general, we can effectively use the dental patterns of extinct mammals to infer their diets. This allows us to develop a picture of the ecological relationships of fossil mammals to each other and to other animal and plant groups at the time they were alive. We can therefore infer something about food chains that existed in times past and compare them with those of modern times to determine the extent of change.

Ecology and Evolution: Domain of the Red Queen

To this point we have been concerned with the factors that influence fossil accumulations and the methods by which scientists gain knowledge about past climate and community structure. But how does this information help us to better understand the evolutionary process? First, the phenotypic expression of an animal is not limited to its structure. We tend to focus on anatomy because it is obvious and because, with fossils, it is all we have to study directly. Yet, it may be something entirely unrelated to structure that determines the ultimate success or failure of a species. As we have seen earlier, speciation rate, a species level property, may play a role. Likewise, climatic change and attendant habitat modification can be a powerful force. Consider the cenograms discussed above. Legendre showed that the cenograms for fossil mammal localities in Oligocene time from southern France after a period of extreme extinction known in Europe as the "Grande Coupure" indicated a rapid change from forested and wet habitats to drier environments. This represents a tie between climate and extinction, which can lead to a significant reorganization of mammalian community structure.

We know that competition occurs in nature, but what is the influence of competition in shaping organismal communities through long periods of time? Do entire communities evolve, or are they just accidental aggregations of species, each with its own separate evolutionary history? How prevalent is *coevolution,* the linked evolutionary history of two or more species? We know from the study of parasite-host relationships and a number of other special interactions that some coevolution occurs in nature. Every species of fig tree, of which there are about 900, is pollinated by a different species of wasp of the Family Agaonidae. Clearly this speaks of a long and bonded history. But can the same be said of predator–prey relationships? After Raymond Lindemann introduced the concept of energy budgets to ecology, the general consensus soon became that the species in an ecosystem interacted in an intricate dance, choreographed by trophic relationships and the Second Law of Thermodynamics. The implication was that species aggregations in ecosystems had been in existence for long periods of time and that the observed foraging patterns among them were intricate, necessary, and highly coevolved. If this is true and if climate plays an important role in evolutionary change of all kinds, we should expect to see wholesale replacement of one group of species in an ecosystem by another if climate changes significantly. Additionally, we might expect to see that morphological change and/or speciation coincides with these climatic episodes. Because of the continued dedication of field paleontologists, we are beginning to obtain a dense fossil record for some geographical regions and time periods, so much so that the analytical techniques described above may soon allow meaningful testing of these and many other important ideas. Let us consider an example.

Meade County in southwestern Kansas is certainly not one's ideal spot for a vacation. It is a flat, generally desolate country, from which most of the native vegetation has been removed by continuous cattle grazing. Other than

Dodge City, about 30 miles to the north, the towns are spotty and small, with an occasional grocery store, a couple gas stations, perhaps a small bar (if the town is not dry, like Meade itself), and a motel or two. The winter wind howls mercilessly from the west, unstoppable as it races across the plains, and the summer heat is intense and dry. It takes a tough cadre of humanity to flourish out there, and that is exactly what one finds. Most of the people are hardworking cattle ranchers. And in one of those incongruous and improbable associations that so often happens in this world, for almost 40 years there has been a wonderful synergism between the people of Meade County and a series of scientists who continue to return to this area because it is one of the most unique paleontological locations in the New World.

Meade County and the surrounding area is sometimes referred to as the Meade Basin, an area in which there was almost continuous deposition of fossiliferous sediments over the past five million years. The deposits were laid down by a combination of small rivers, streams, ponds, and artesian systems. For more than 30 years this area was studied by the late C. W. Hibbard and his students from, first, the University of Kansas and later at Michigan. In addition to discovering many fossiliferous localities and naming many new kinds of animals, mostly small mammals, Hibbard's greatest contribution to science will undoubtedly be seen as the introduction of screenwashing to paleontology. Before his work, paleontologists were concerned only with large impressive fossil animals, but Hibbard demonstrated, initially just using window screen, that one could add many species to each fossil locality by sieving the dirt and picking out the remains of small animals, mostly vertebrates (fishes, amphibians, reptiles, birds, mammals). He also showed that the past environments of fossil localities could be better interpreted by examining these small mammals. (From this initial work also came the realization that small mammal remains, so ubiquitous in many localities over the past five million years, offered tremendous potential to study both the tempo and mode of evolution. As with all great innovations, the spinoffs are often unpredictable and at least equally as valuable as the initial discovery.) Work continues in the Meade Basin. A team of faculty and students from Murray State, the University of Colorado at Boulder, and the National Museum of Natural History in Madrid celebrate Hibbard's legacy by prospecting for new localities with the next generation of kind and considerate Meade County ranchers, as virtually all the fossil localities are on private land. Nowhere else in the New World can one find so many localities spanning such a long period of time in close proximity. Only here is it possible to actually trace the history of an ecological community through millions of years and see the transition to the modern community of the same region. We have specifically been examining the history of the rodent fauna (mice, voles, squirrels, etc.) because they are so well represented here as fossils. And our preliminary analyses suggest that ecological associations do not remain together for long periods of time (in the geological sense). Working predominantly with the fossil materials Hibbard had collected over the years, we have shown that the rodent species groups change in composition regularly and continuously. And it is not that they are evolving from one kind into another; they are replaced, apparently through the interplay of competitive interactions and habitat modification. In fact, the modification in their numbers over time appears to conform to a principle proposed by a scientist at the University of Chicago more than 20 years ago.

In 1973 the evolutionary biologist Leigh Van Valen suggested that related species in any group go extinct at a rate that is constant for each group. This is sometimes known as the "Law of Constant Extinction" or "Van Valen's Law." It is visually represented by a plot of the log of the percent surviving of a group against absolute time (**Fig. 3.8**). In ecology, this type of representation is known as a "survivorship" plot and is basically the same approach as is used by actuarians working for insurance companies figuring out everybody's insurance rates (e.g., how many females born in 1944 who smoke three packs of cigarettes die each year in car crashes). To explain this pattern in the fossil record, Van Valen suggested that species are constantly competing with others in their ecological niche (or adaptive zone) and also trying to keep up with abiotic changes in their environment, such as increased cold or aridity. He called this the "Red Queen" hypothesis, in honor of the queen's comment from Lewis Carroll's *Through the Looking Glass:* "Now here you see, it takes all the running you can do, to keep in the same place." In Van Valen's world, species continually compete for resources as the environment around them changes. They try to make adjustments, but eventually the combination of new competitors and environmental modification wears them down. Despite the concordance of our data with Van Valen's model, there are some possible interesting fine points yet to be established. For example, it seems possible that larger predators and their prey may be more highly coevolved because larger animals appear to have longer species life spans than smaller ones. Carnivores like weasels and foxes must be very flexible and opportunistic, whereas wolves and large cats might be more tightly limited by their prey and therefore might attune their behavioral strategies more to specific prey. However, as history also shows, this is not a particularly compelling strategy either, because all the saber-toothed cats that have lived worldwide are no longer with us, clearly the result of the extinction of the specific large prey items on which they fed. The Law of Constant Extinction and its attendant maid, the Red Queen, tell a sobering story and leave us with a somewhat clear evolutionary moral: Extinction is inescapable under natural conditions. Will humans succumb as well? We seem to have wiped out the competition, so maybe there is a chance we will survive.

Figure 3.8 Survivorship plots from work of Leigh Van Valen, showing constant rate of extinction among a variety of vertebrate groups over millions of years. (From L. Van Valen, 1973. A new evolutionary law. *Evolutionary Theory* 1:1–30.)

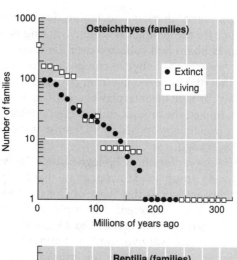

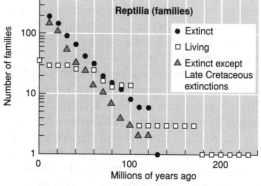

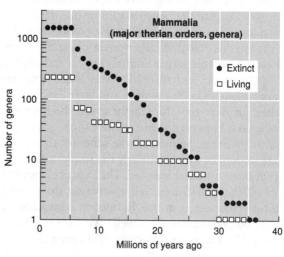

Suggested Readings

Brown, J. H. 1995. *Macroecology.* University of Chicago Press, Chicago.

Diamond, J. and T. J. Case (eds.). 1986. *Community Ecology.* Harper & Row, New York.

Eldredge, N. and J. Cracraft. 1980. *Phylogenetic Patterns and the Evolutionary Process.* Columbia University Press, New York.

Graham, R. W., H. A. Semken, Jr., and M. A. Graham (eds.). 1987. *Late Quaternary Mammalian Biogeography and Environments of the Great Plains and Prairies.* Illinois State Museum, Springfield. Scientific Papers 22.

Grayson, D. K. 1993. *The Desert's Past: A Natural Prehistory of the Great Basin.* Smithsonian Institution Press, Washington, DC.

Hennig, W. 1966. *Phylogenetic Systematics.* University of Illinois Press, Urbana.

Hillis, D. M., C. Moritz, and B. K. Mable (eds.). 1996. *Molecular Systematics.* Sinauer, Sunderland, Massachusetts.

Peters, R. H. 1983. *The Ecological Implications of Body Size.* Cambridge University Press, New York.

Pielou, E. C. 1991. *After the Ice Age: The Return of Life to Glaciated North America.* Chicago University Press, Chicago.

Ricklefs, R. E. and D. Schluter. 1993. *Species Diversity in Ecological Communities: Historical and Geographical Perspectives.* University of Chicago Press, Chicago.

Rosenzweig, M. L. 1995. *Species Diversity in Space and Time.* Cambridge University Press, New York.

Strong, D. R., D. Simberloff, L. G. Abele, and A. B. Thistle (eds.). 1982. *Ecological Communities: Conceptual Issues and the Evidence.* Princeton University Press, Princeton, New Jersey.

4

Origins

"With the concept of evolution life's patterns and diversity become meaningful and comprehensible; without it there is only magic and superstition."

Are we Martians? As ludicrous as this sounds, it may be closer to the truth than we care to admit. In 1996 David McKay and colleagues reported finding the fossilized remains of microbes in a four billion year old meteorite from Mars. The rock had fallen onto the ice in Antarctica. This raises the possibility (another hypothesis) that life existed on at least two planets in the solar system, and if life originated first on Mars, it may have seeded the Earth. This idea, called the *panspermia* hypothesis, could of course also work in reverse. Life could have arisen on Earth and been transported to Mars. Maybe there was a continuous exchange, which has given rise to the colloquial "ping-pong ball" hypothesis of multiple seedings in both directions. As fanciful as these musings seem, the new Martian evidence indicates we now have to take these ideas seriously. Before we delve into this matter any further, let us take a look at the presumed early atmosphere of Earth and the experiments that led some scientists to suggest that life can arise spontaneously from nonliving elements under proper conditions. This is followed by a brief review of the fossil record of early life on Earth, further evidence in support of evolutionary theory.

Origin of Life

Early Atmosphere and Seas

The Earth is about 4.5 to 4.6 billion years (Gy) old. The entire time period from about 4.5 to 0.59 billion years ago (Ga) is sometimes referred to as the Cryptozoic Eon and the period from 0.59 Ga to the present as the Phanerozoic Eon. The Cryptozoic Eon is further broken into the Archean Era (4.5 to 2.5 Ga) and the Proterozoic Era (2.5 to 0.59 Ga). According to the latest assessments of the Earth's formation, with the exception of the occasional huge asteroid impact that would have had a devastating effect on any living things, the atmosphere could have been conducive for life from about 4.4 Gy onward, when surface temperatures had reached a level below 100°C. But this early period is a black box now, and because the earliest continental sediments are dated at about 3.8 Ga, it may be that the Earth's surface was just too unstable for life's appearance until much before that time (**Fig. 4.1**). According to Sherwood Chang of the National Aeronautics and Space Administration, the early atmosphere was fueled by volcanism and impacts. It is currently believed that the earliest atmosphere was composed mostly of carbon dioxide, carbon monoxide, nitrogen, water vapor, methane, and some sulfur-containing gases. Because the sun was about 25% to 30% dimmer in those days, a considerable layer of greenhouse gases was necessary to maintain moderate temperatures at the Earth's surface. Carbon dioxide may have been present at 100 to 1000 times the concentration it is today. Although there is some debate on this issue, most scientists have tentatively concluded there was

Figure 4.1 Geologic time chart and important events in the early history of the Earth. (Adapted from N. Campbell, L. Mitchell, and J. Reece, 2000. *Biology: Concepts and Connections,* 3rd ed. San Francisco: Benjamin Cummings. p. 319.)

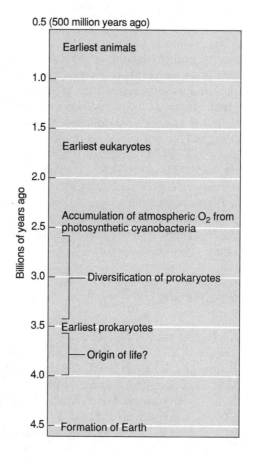

little free oxygen in the earliest atmosphere. The arguments are fairly detailed, but suffice it to say there was not enough bioproduction of oxygen in the early seas to counter the chemicals that could combine with the oxygen that was produced (this process is referred to as *oxidation*). There is no conclusive evidence pointing to an exact time when atmospheric oxygen levels reached those comparable with modern time, but it is likely to have taken place between about 1.5 and 0.6 Ga, when the stromatolite cyanobacterial mats were still common and multicellular marine algae had also appeared.

The Earth's surface region is known as the *lithosphere*. From very early in the Earth's history it is likely the lithosphere was composed primarily of seawater, originating either from water vapor produced by volcanic outgassing or, amazingly, from constant bombardment by comets, which are predominantly composed of frozen water. The latter theory is becoming more likely as we continue to learn about the early formation of solar systems (more on this later in the chapter). Land, referred to as "microcontinental blocks" by some authors, probably made up less than 5% of the Earth's surface and was very likely restricted to volcanic islands sitting on mantle "hotspots," much as the Hawaiian Islands do today. The composition of this early Archean sea was the result of interactions between the dynamic mantle below and what little cooled and hardened crust existed. Some investigators believe the early oceans could have been highly acidic, but this is debatable. They were certainly relatively sterile, because there was no nutrient runoff from land as there is today. Donald Lowe of Stanford University refers to the Archean seas as a "nutrient-starved desert." A significant change in oceanic dynamics and composition occurred as more land appeared in the Proterozoic. This is reflected by different types of sediments indicating erosion from land and higher levels of bioproductivity. Hydrothermal vents were probably common in deep seas trenches, and it is here, say some scientists, that the first life may have arisen.

Life's Assembly Rules

Phylogenetic Considerations

If interpretations of the early Earth environment are correct, the first living things on this planet must have been anaerobic (**Fig. 4.2**). This speculation is actually consistent both with the most recent comparative phylogenetic studies of living things and the modern environments within which we find the most primitive living organisms. Recent studies of the "tree of life" disclose that the early evolution of living things was much more complex than represented by the now-classic "five kingdom" classification (Bacteria, Protista, Animalia, Fungi, Plantae). For example, as shown by N. Iwabe and colleagues, the split between the Archaebacteria and the Eubacteria is more pronounced in certain ways than the differences between plants and animals or fungi.

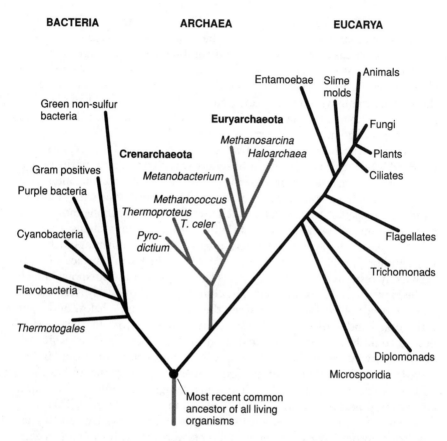

Figure 4.2 Three Domain (or Kingdom) model for life's phylogeny. In this model, one of the ancient groups of bactera-like organisms (the Archaea) is as different from other bacteria-like organisms (Bacteria) as it is from all other living things. (Adapted from *Evolutionary Analysis,* 2e, by Freeman/Herron, copyright 2001. Reprinted by permission of Pearson Education, Inc., Upper Saddle River, NJ.)

Therefore the three broad divisions of life, perhaps called domains, can be called the Eubacteria, the Archaea (the Archaebacteria), and the Eucarya (the eukaryotes). RNA comparisons show that thermophilic members of both the Eubacteria and the Archaea form the initial branches from life's universal phylogeny. These results point to a chemosynthetic origin of life in a hot environment; possibly hydrothermal vents in the ancient oceans. The heated habitats in which we find thermophilic prokaryotes today are characteristically anaerobic or only negligibly aerobic. Nevertheless, a recent surprise pointed out by Karl Stetter of Regensburg University in Germany is the presence of very primitive thermophiles that require trace amounts of oxygen to survive. It may be that both aerobic and anaerobic prokaryotes have been around since early Archaean times, but aerobic metabolism did not expand in organisms until critical levels of oxygen were attained in the atmosphere during the late Proterozoic. With this phylogenetic setting in mind, let us take a look at the chemical conditions under which the first cells might have appeared and some experiments that have been conducted to test these ideas.

The Broth of Life

As Juan Oro of the University of Houston notes, we live in a universe in which, other than the noble gases He and Ne, the most abundant elements are hydrogen, oxygen, carbon, and nitrogen, exactly the makeup of organic matter. Carbon compounds and radicals, including the following, have been recorded from comets: C_2, C, CH, CO, CS, HCN, and CH_3CN. A large number of amino acids, carboxylic acids, purines, pyrimidines, and hydrocarbons have been recovered from small meteorites. In 1969 a meteorite broke up over the town of Murchison, Australia. Scientists recovered a variety of amino acids from the interior of some of the pieces, including glycine, alanine, glutamic acid, and valine. It is therefore not unreasonable to assume, at least temporarily, that organic (carbon-based) chemistry dominates in the cosmos. Simple combinations of carbon, nitrogen, and hydrogen form naturally as the byproduct of stellar fusion and are known to be relatively stable at temperatures as high as $6000°$ K. These molecules are constantly being shed into space, where they may gain complexity as they adhere to dust particles and other substrates and react with other molecules. Many of the molecules mentioned above have also been detected in interstellar clouds.

During the early phases of the formation of our solar system, interstellar clouds and debris were coalescing to form the planets. During this period there were tremendous numbers of collisions, often between objects of massive size. Our Moon, for example, likely resulted when an object the size of Mars collided with the "protoearth." It sent a chunk of debris outward, to eventually be captured in the Earth's gravitational field as the Moon, and it likely vaporized all the water and sent into space virtually all the organic volatile substances that could have developed into life. The Earth then became, before about 4.4 Ga, a dry barren sphere. So where then did all the water and early chemicals come from? Very likely from outer space. A number of individuals, including Juan Oro, have suggested the source of all the Earth's organic compounds, as well as its water, was from interstellar sources, most likely comets. As Oro notes, support for this model is being provided indirectly from astronomical observations made of other solar systems, one of which is named Beta Pictoris. This system is believed to be in the early stages of planetary formation, and astronomers interpret strong variations in certain spectral lines detected by an infrared telescope to be the result of comet-like bodies falling toward Beta Pictoris at a rate of about 10 to 100 impacts per year. Depending on the size of these bodies, that rate would be sufficient to provide all of the Earth's water and its atmosphere.

In addition to extraterrestrial bombardment by objects rich in water and carbon compounds, modern experiments have shown that the building blocks for organic life can be synthesized very easily and quickly by introducing energy into simple combinations of abiotic chemicals presumed to be present in the Earth's early atmosphere. The Russian Alexander Oparin first suggested the abiotic synthesis of biotic life, and in the 1950s Stanley Miller at the University of Chicago successfully synthesized amino acids and other biochemicals in Harold Urey's laboratory by running a spark into a combination of methane, ammonia, and hydrogen (**Fig. 4.3**). Ultraviolet light can act on simple combinations to create biochemicals, and virtually all the precursors for organisms, including nucleic acids and adenosine, have now been synthesized under laboratory conditions. Available evidence suggests that the early atmosphere of the Earth was primarily carbon dioxide rather than the methane-ammonia mixture used in Miller's experiments. This would make it much more difficult to manufacture organic molecules. However, the ammonia and methane in the prebiotic atmosphere may have been provided by the bombardment of carbon-rich comets and asteroids. The "primeval soup" would have been rich indeed. How then were these chemicals organized into cells?

Plasma membranes that surround cells are essentially double layers of lipids (fats), and it has been shown in a number of laboratory experiments that various lipids can self-assemble into membranes and cellular-like structures called *liposomes*. These membranes have even been shown to be semipermeable to certain ions and weak bases, like membranes in living cells. The first prokaryotic cells must have resulted from the assembly of appropriate molecules into these cell-like structures. We do not know when early protocells crossed the line from nonliving to living, but given about a half billion years of natural experimentation (from about 4.5 to 3.8 Ga, when life is at least indirectly identified on Earth) with the primeval broth being challenged by every conceivable source of energy, including heat, lightning, extraterrestrial bombardment by comets and meteorites, ultraviolet light, and cosmic radiation, it does not appear too surprising; it is just a minor rather than a major miracle.

And this brings us back to Mars. Given what we now know about the interstellar travel of carbon compounds, it would come as no great shock to find such compounds on a four billion year old meteorite originating from Mars—just another source for the early Earth kettle. What makes this new discovery so fascinating is that the meteorite might have carried something that was already alive. This finding does not affect our scenarios for the first appearance of life in the solar system, just *where* it first appeared. Perhaps

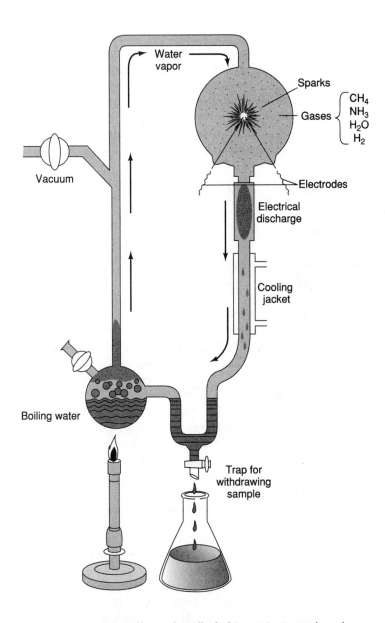

Figure 4.3 The laboratory apparatus used by Stanley Miller for his experiments on the early Earth atmosphere. The simple mixture of water, carbon dioxide, nitrogen, and hydrogen produced a variety of organic compounds (e.g., alanine, succinic acid, urea) when charged with an electric spark. Given the complex organic molecules present on meteorites that have hit the Earth in modern time, it is likely the early Earth 4.5 Ga was one large chemistry experiment that went on until life arose.

life first arose on Mars and then seeded the primitive Earth seas. Or maybe it was vice versa, or there could have been numerous exchanges of near-living and living cells. Given how easy it is to synthesize all the building blocks for life, nobody is yet willing to take seriously the hypothesis that the Earth (and maybe also Mars) was purposely seeded with simple living cells by an extraterrestrial civilization, but it has become admirably clear that, in addition to being starstuff ourselves, the cosmos continues to have a profound influence on the dynamics of our beautiful blue planet.

The Fossil Record of Early Life

The Earliest Fossils: 3.4 to 0.61 Ga

A cursory review of the fossil record shows that the Cambrian Period of the Paleozoic Era is the first period of time during which complex forms of advanced life are seen in abundance. An example of this is the 0.530 My or 0.53-Gy Burgess Shale fauna of British Columbia, made famous in Stephen Jay Gould's book, *Wonderful Life*. Before the Cambrian (before 0.59 Gy), the records of life are spotty. The fossil record of ancient life is not as bountiful as we would like primarily because few of the most ancient rocks are still with us. Most have been eroded away. However, those rocks and fossils that are around continue to confirm the predictions of evolutionary theory. If life's full diversity evolved by evolutionary processes, then we should see a tree of life unfolding in a predictable manner, with microbes populating the earliest sediments and more complex organisms appearing later. This is not to imply any kind of "progress," directed or otherwise, except to the extent that certain structures and functions must logically precede others. For example, if vertebrates arose in water, then we should expect to go back to a time when there were only aquatic vertebrates. The first amphibians should appear in younger sediments than the first fishes, and so on. Dinosaurs, as advanced reptiles, should appear after the first amphibians if, as we suppose, reptiles evolved from an amphibian ancestor. The creationist model would be supported if none of these patterns was observed or if dinosaurs, humans, and the first amphibians were recovered from contemporaneous deposits.

The oldest fossils of living organisms come from the Apex Chert of the Warrawoona Group of rocks in Western Australia (**Fig. 4.4**). Described by J. William Shopf, these 3.4 Gy old fossils appear to be filaments of a cyanobacterium (formerly known as the "blue-green algae"). They belong to the genus *Primaevifilum*. Although the identification of the species as a photosynthetic microbe is tentative, if correct it indicates that life arose considerably earlier, because cyanobacteria are rather advanced over the ideal first microbe. But there must be a limit on when life could have arisen, because if we go back

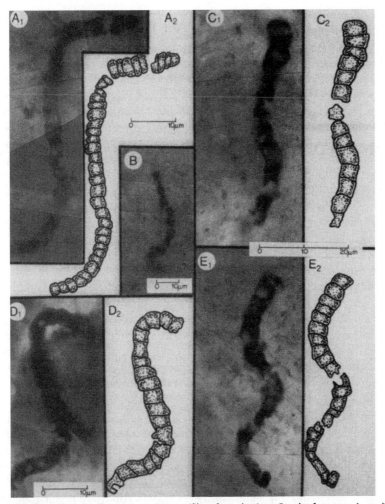

Figure 4.4 The earliest fossil organisms, *Primaevifilum*, from the Apex Basalt of western Australia, at approximately 3.4 Ga. This organism has been interpreted as a colonial cyanobacterium, which implies that simpler life evolved considerably earlier. (From J.W. Schopf, 1992. The oldest fossils and what they mean. In J.W. Schopf, ed. *Major Events in the History of Life*. Sudbury, MA: Jones and Bartlett. p. 43.)

far enough the Earth was a molten mass, consolidating from interstellar debris. The earliest rocks on Earth indicating a cooling crust are about 4.2 to 4.3 Gy old. We also know that surface water existed by about 3.8 Ga, because there are sedimentary rocks of that age in Greenland. Indirect evidence from carbon isotope residues trapped within apatite crystals suggests that some of the constituents of these early Greenland rocks were formed by organisms, although this is currently controversial. Apatite is a mineral formed from calcium phosphate, the same stuff that makes up our teeth and bones. These dates do not preclude the origin of life much earlier, but we lack the evidence to determine this now.

Many Archean fossils are represented by *stromatolites,* pillar-like structures composed of lamina formed by cyanobacteria strands (**Fig. 4.5**). Amazingly, such structures still exist today in warm shallow bays off the coast of Australia. When sectioned, they have a distinctive anatomy and have been recognized from many localities all over the world. So stromatolites appear to be the oldest complex structures on Earth. They are well represented in deposits greater than 3.3 Ga in the Onverwacht Group of rocks from South Africa and become widespread during the Proterozoic Era. They diminish greatly in abundance at the beginning of the Phanerozoic.

Unequivocal remains of eukaryotic fossils, those with nuclear membranes, come from 2.1-Ga banded iron deposits from a mine in Michigan. Named *Grypania spiralis* by T.-M. Han and B. Runnegar, they are considered to be algae. If correctly interpreted, Bruce Runnegar has suggested the origin of eukaryotes must be much earlier. In sediments younger than 2.0 Ga, both unicellular and multicellular eukaryotes become common and include phytoplankton, microscopic red algae, calcareous algae, complex protists, larger algae, and what has been referred to as the "Ediacara fauna," a mixed bag of simple and complex creatures that existed toward the end of the Proterozoic Era. The time during which the Ediacara fauna existed, from 610 to 550 Ma, is known as the Vendian Period and is worth a cursory examination.

Vendian Period

Mikhail Fedonkin describes the later Proterozoic as a time of radical change in global oceanic ecosystems. The continental plates developed to a great degree beginning about 2.6 Ga, which saw an expansion of shallow marine habitats. The levels of atmospheric oxygen increased due to the expansion of photosynthetic organisms (e.g., the stromatolites and, later, true algae) and levels of carbon dioxide significantly decreased. Alternate periods of glaciation and warming have been recorded since about 0.85 Ga. There was of course a steady increase in the sun's output. All these events led to significant modifications in physical and chemical properties of the oceans, which set the scene for the appearance of more complex organisms and communities.

The Varanger glacial period marks the advent of the Vendian all over the world. Shallow seas that had extended into continental regions (known as *epicontinental* seas) retreated, and thus the shallower and presumably more productive regions of the ocean became restricted to narrow belts near the deeper bodies of water. The Vendian is characterized as a period of such glacial regressions followed by warm transgression periods when the seas once again rose. Throughout the Vendian we see numerous localities all over the

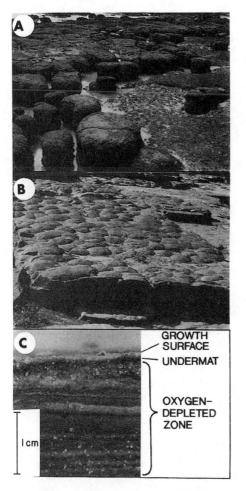

Figure 4.5 Stromatolites, mats of cyanobacteria forming mushroom-like clumps, alive today in Shark Bay, western Australia. Mats of cyanobacteria were common during the Proterozoic eon. (From J.W. Schopf, 1992. The oldest fossils and what they mean. In J.W. Schopf, ed. *Major Events in the History of Life.* Sudbury, MA: Jones and Bartlett. p. 34.)

world with a diverse fauna of algae, invertebrates, and so-called trace fossils (physical evidence of organisms, such as raised burrows or trails). The great question is to what extent Vendian organisms gave rise to the explosive radiation of animals at the beginning of the Phanerozoic (the "Cambrian explosion").

Many Vendian organisms were very large, giant when compared with the small creatures of the earliest Cambrian. Some of the Ediacara taxa were in excess of 1 meter in size, as opposed to a few millimeters for most Cambrian animals. The absence of internal skeletons is a striking feature of Vendian invertebrates, though it is not so surprising when we consider that about 70% of modern marine species can be considered "soft bodied." Perhaps the most

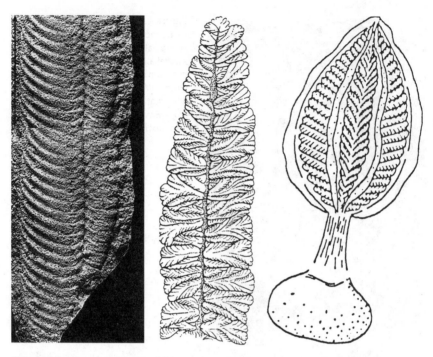

Figure 4.6 Examples of the Vendobionta, unusual and enigmatic organisms of the "Ediacara fauna" from the Vendian period, about 600 Ma. (Left: Photograph courtesy of M.A. Fedonkin. Middle: Drawing courtesy of A. Seilacher. Right: from R.J.F. Jenkins, 1985. The enigmatic Ediacaran (late Precambian) genus *Rangea* and related forms. *Paleobiology* 11:336–355.)

important idea suggested by this observation is that the skeletons of Cambrian organisms likely originated very quickly at the Phanerozoic boundary; they did not have a long pre-Cambrian history.

Another interesting feature of Vendian organisms, virtually the entire Ediacara fauna, is that they were morphologically flattened into leaf or frond-like forms. This does not seem to be the result of deformation by the sediments. Fedonkin accepts the form of these organisms as natural and suggests they were specially adapted to the flattened surface of stromatolite mats. It may even be that eukaryotes arose on these mats. The German paleontologist Adolf Seilacher considers the flat Ediacaran organisms to represent an entirely unique and phylogenetically separate organismal group, sometimes referred to as the "Vendobionta" (**Fig. 4.6**). He claims these organisms did not possess internal digestive or circulatory systems similar to later complex invertebrates, which might in part explain their large size and flattened quilted surfaces, because they needed this larger surface area/volume ratio to exchange gases and nutrients necessary for survival. To Seilacher, the Vendobionta are not even multicellular and not metazoans. Fedonkin agrees that the Vendobionta were unusual and unrelated to modern organisms, but he disagrees with Seilacher that they were not metazoans, and he has reported digestive structures in a number of them. Regardless, it appears there was an early radiation of unusual forms, dominant during the later Proterozoic, that became extinct and are unrelated to animals that appeared during the early Phanerozoic. It is a classic example of explosive adaptive radiation of a clade followed by its extinction.

Because those species that were likely ancestral to Phanerozoic organisms were mostly soft bodied, their remains are rare in Vendian sediments, and it is not clear which taxa might have given rise to the Cambrian faunas. Nevertheless, Fedonkin and others point to the medusa-like Cyclozoa, schyphozoan-like *Conomedusites*, which has tentacles and a thin stem attachment, and a variety of organisms that has at various times been considered arthropods, such as *Mialsemia, Vendia, Bomakellia, Dickinsonia,* and *Vendomia,* all of which are from Vendian rocks along the White Sea coast in Russia. We cannot say for certain which, if any, of these forms are ancestral to Phanerozoic taxa, but we can see that they are generally of a morphologically simpler grade and consistent with the pattern expected and predicted by evolutionary theory.

Cambrian Revolution

I concentrate here on the famous Burgess Shale fauna of British Columbia because it includes representatives of the vast majority of groups represented around the world by early Cambrian time. Another equally impressive early Cambrian fauna is a recently discovered aggregation known as the Chengjian fauna from Yunnan, China. There is no doubt that the early Phanerozoic is a focal time in the evolution of metazoan life. After all the smoke clears from the debates and studies going on about organisms of this time period, most likely we shall see that all modern species can be traced back to organisms that are known from this time. This in spite of the likelihood, as we saw in the Vendian, that many unusual animal groups, unrelated to modern organisms, also existed side by side with these ancestral taxa.

Charles Doolittle Walcott discovered the Burgess Shale fauna in 1909. At that time Walcott was Secretary (the equivalent of Director) of the Smithsonian Institute in Washington and one of the world's leading authorities on extinct invertebrates, especially trilobites. The locality from which the fauna is derived sits high in the mountains of British Columbia, near Banff in Yoho National Park. To get to the site in Walcott's day required a difficult journey by horseback. Today we understand that the shales represent a mudslide that occurred 530 million years ago on a shelf below a reef that stood in shallow marine waters. The organisms in the shales were entombed instantaneously, resulting in a treasure trove of both soft- and hard-bodied fossils, known in paleontological vernacular as a "lagerstätten." Walcott described the fauna in a series of detailed papers, and he suggested that the animals fit well into known animal groups, including the arthropods, trilobites, cnidarians (coelenterates), and a variety of other modern taxa. The fauna was considered interesting only in that it represented the sudden appearance of so many taxa so quickly in the fossil record and helped to form the notion of a "Cambrian explosion" of advanced metazoan life. In this sense, the Cambrian still occupies that unique position. But Walcott's taxonomy of the Burgess Shale animals, it turns out, was mostly wrong.

In the 1960s, the Canadian government asked Des Collins, a geologist, to make additional collections of the Burgess Shale fauna. Subsequently, Harry Whittington, a professor at Cambridge, was invested with the task of restudying the new materials as well as the old. He and two of his students, Simon Conway Morris and Derek Briggs, took on this prodigious task, and over the past 30 years they have been involved in this work and have published monographs that amazed the scientific world and formed the basis of a popular account written by Stephen Jay Gould of Harvard University, titled *Wonderful Life*. Although the book has been criticized and some of the conclusions since overturned, it undoubtedly brought more attention to the Burgess Shale animals and the early evolution of life than all other scientific publications on the locality combined.

The most dramatic discovery of the Burgess Shale fauna is that it was composed of a tremendous diversity of body forms, many of which do not seem to fit within known animal phyla (**Fig. 4.7**). They can be termed bizarre for lack of a better word. *Wiwaxia, Opabinia,* and *Amiskwia* look like concoctions made for a children's cartoon series; there are animals with five eyes (so much for complete bilateral symmetry). *Hallucigenia* was thought to also represent one of these weird creatures, but more complete material from a Chinese locality shows that it was likely a strange onychophoran worm. What had originally been interpreted as dorsal spines were later shown to be "tube feet"; *Hallucigenia* had been reconstructed upside down. Besides these strange beasts, a number of species are present that do fit into modern phyla and likely represent at least general body plan ancestors for living animals. Among them is our likely ancestor, the early chordate *Pikaia,* an unobtrusive free-swimming slug-like creature. But why did Walcott misinterpret all these animals? Gould believes Walcott was brainwashed by three prevailing influences: that life increased in diversity and complexity in a "cone of diversity," that evolution unfolded gradually, and that there was progress to evolution, culminating with humans. Gould makes a very good argument that Walcott simply could not envision a tremendous and geologically rapid radiation of

Figure 4.7 Reconstructions of *Opabinia, Wiwaxia, Anomalocaris,* and *Pikaia* from the famous early Cambrian Burgess Shale fauna of British Columbia, about 530 Ma. (S.J. Gould, 1989. *Wonderful Life. The Burgess Shale and the Nature of History*. New York: Norton.)

organisms in the Cambrian, the vast majority of which would later become extinct. The Aristotelian concept of the "ladder of progress" was also an underlying theme of much scientific work of Walcott's day, as it still is in Christian fundamentalism. This idea proposes that life evolves from the simple to the complex and (perhaps inadvertently) implies that humans occupy the highest point on the ladder and are therefore closest to "perfection." So, many scientists saw life as a slow and methodical march of intermediate forms from simple microbes to humans.

We know today that the appearance of organismal groups is often episodic, with explosive radiation often followed by periods of, sometimes massive, extinction. Most of the animals from Burgess Shale time did not survive to leave descendants to modern time. But far from being atypical, this process seems to be common, and the probability of long-term survival amounts often more to a set of chance phenomena than it does to any form of direction or inherent progress. Gould describes this as "contingency." That is, the success or direction evolution takes is contingent (dependent) on a series of unlikely events that preceded it. If we were able to rewind the tape of life and replay it, the results almost certainly would be different. Slight changes would result in tremendously different results. For instance, if an asteroid had not hit the Earth 65 million years ago, the dinosaurs would likely not have become extinct, the mammals would not have become the dominant vertebrates of Cenozoic time, and if intelligent animals came to exist at all, they would have had scales instead of fur.

The Cambrian revolution saw the first filling of the ecological barrel and set the scene for the eventual appearance of all modern animals. It is exactly what we would expect of an intermediate set of organisms between an earlier time period composed exclusively of prokaryotes and simple eukaryotes and later periods with more advanced animals such as the vertebrates. There is no fossil evidence here to contradict Darwin's mighty theory. With the concept of evolution, life's patterns and diversity become meaningful and comprehensible; without it there is only magic and superstition.

Suggested Readings

Achenbach-Richter, L., K. O. Stetter, and C. R. Woese. 1987. A possible biochemical missing link among archaebacteria. *Nature* 327:348-349.

Barghoorn, E. S. and J. W. Schopf. 1966. Microorganisms three billion years old from the Precambrian of South Africa. *Science* 152:758-763.

Bengston, S. (ed.). 1994. *Early Life on Earth.* Columbia University Press, New York.

Benlow, A. and A. J. Meadows. 1977. The formation of the atmospheres of the terrestrial planets by impact. *Astrophysics and Space Science* 46:293–300.

Blankenship, R. 1992. Origin and early evolution of photosynthesis. *Photosynthesis Research* 33:91–111.

Briggs, D. E. G. and R. A. Fortey. 1989. The early radiation and relationships of the major arthropod groups. *Science* 246:241–243.

Briggs, D. E. G., R. A. Fortey, and M. A. Wills. 1992. Morphological disparity in the Cambrian. *Science* 256:1670–1673.

Conway Morris, S. 1989. Burgess Shale faunas and the Cambrian explosion. *Science* 246:339–346.

Conway Morris, S. 1993. The fossil record and the early evolution of the Metazoa. *Nature* 361:219–225.

Gould, S. J. 1989. *Wonderful Life.* Norton, New York.

Iwabe, N., K. Kuma, M. Hasegawa, S. Osawa, and T. Miyata. 1989. Evolutionary relationship of archaebacteria, eubacteria, and eukaryotes inferred from phylogenetic trees of duplicated genes. *Proceedings of the National Academy of Sciences USA* 86:9355–9359.

Joyce, G. F. 1989. RNA evolution and the origins of life. *Nature* 334:564.

Kabnick, K. S. and D. A. Peattie. 1991. *Giardia,* a missing link between prokaryotes and eukaryotes. *American Scientist* 79:34–43.

Kasting, J. F. 1993. Earth's earliest atmosphere. *Science* 259:920–926.

Lazcano, A. and S. L. Miller. 1996. The origin and early evolution of life: prebiotic chemistry, the pre-RNA world, and time. *Cell* 85:793–798.

Margulis, L. 1993. *Symbiosis in Cell Evolution: Microbial Communities in the Archean and Proterozoic Eons.* Freeman, New York.

McKay, D. S., et al.. 1996. Search for past life on Mars: possible relic biogenic activity in Martian meteorite ALH84001. *Science* 273:924-930.

McMenamin, M. A. S. and D. I. S. McMenamin. 1990. *The Emergence of Animals.* Columbia University Press, New York.

Miller, S. J. and J. L. Bada. 1988. Submarine hot springs and the origin of life. *Nature* 334:609–611.

Miller, S. J., H. C. Urey, and J. Oro. 1976. Origin of organic compounds on the primitive Earth and in meteorites. *Journal of Molecular Evolution* 9:59–72.

Mojzsis, S. J., G. Arrhenius, K. D. McKeegan, T. M. Harrison, A. P. Nutman, and C. R. L. Friend. 1996. Evidence for life on Earth before 3800 million years ago. *Nature* 384:55–59.

Oberbeck, V. R. and G. Fogelman. 1989. Impacts and the origin of life. *Nature* 339:434.

Oro, J., S. L. Miller, and A. Lazcano. 1990. The origin and early evolution of life on Earth. *Annual Review of Earth and Planetary Sciences* 18:317–356.

Oro, J., T. Mills, and A. Lazcano. 1992. Comets and the formation of biochemical compounds on the primitive earth-a review. *Origins of Life and Evolution of the Biosphere* 21:267–277.

Sagan, C. and G. Mullen. 1972. Earth and Mars: evolution of atmospheres and surface temperatures. *Science* 177:52–56.

Schopf, J. W. 1993. Microfossils of the Early Archean Apex Chert: new evidence of the antiquity of life. *Science* 260:640–646.

Schopf, J. W. 1994. The early evolution of life: solution to Darwin's dilemma. *Trends in Ecology and Evolution* 9:375–378.

Schopf, J. W. and C. Klein (eds.). 1992. *The Proterozoic Biosphere.* Cambridge University Press, Cambridge.

Vermeij, G. 1990. The origin of skeletons. *Palaios* 4:585–589.

Whittington, H. B. 1985. *The Burgess Shale.* Yale University Press, New Haven.

5

Pedestrian Whales

". . . while in the earlier geological strata there are found fossils of monsters now almost entirely extinct; the subsequent relics discovered in what are called the Tertiary formations seem the connecting, or at any rate intercepted links, between the antechronical creatures, and those whose remote posterity are said to have entered the Ark."

This interesting combined evolutionary and fundamentalist statement appears in Herman Melville's *Moby Dick*. In a quaint section on fossil whales, Ishmael, Melville's hero, waxes intelligently about the then known fossil history of whales but also clearly separates this fossil history from the time of "Man" and the Biblical flood, after which modern time can be said to have begun. Melville's scholarly knowledge can be seen in the following statement:

> *All the Fossil Whales hitherto discovered belong to the Tertiary period, which is the last preceding the superficial formations. And though none of them precisely answer to any known species of the present time, they are yet sufficiently akin to them in general respects, to justify their taking ranks as Cetacean fossils.*

On the other hand, Melville does not seem to be able to place these fossilized remains into a common context with humans and has a curious perspective on time itself, as we can see by Ishmael's continued commentary:

> *When I stand among these mighty Leviathan skeletons, skulls, tusks, jaws, rigs, and vertebrae, all characterized by partial resemblances to the existing breeds of sea-monsters; but at the same time bearing on the other hand similar affinities to the annihilated antechronical Leviathans, their incalculable seniors; I am, by a flood, borne back to that wondrous period, ere time itself can be said to have begun; for time began with man.*

There is an element of William Paley's natural theology here, and it is clear from a set of quotes on whales appended to *Moby Dick* that Melville was aware of Paley's writings. We now know that many of the specific fossil whales of which Melville wrote lived more than 40 million years ago (Ma) and, with additional fossil material, some only recently discovered, they provide one of the best documented long-scale examples of evolutionary links in the fossil record.

George Gaylord Simpson, one of the premiere paleontologists of the twentieth century, once remarked that whales were so unusual that "There is no place for them in a *scala naturae.*" Then, about 50 years ago, Leigh Van Valen made the outrageous suggestion that whales were likely related to hooved mammals such as the Artiodactyla (even-toed ungulates: deer, pigs, buffalo, camels, etc.), this despite the carnivorous habits of many living whales. Many sources of information, particularly from molecular sequence data in modern mammals and the fossil record, now support Van Valen's early idea, though the specific ancestral group remains in question. Van Valen originally proposed that whales were descended from an extinct group of hooved carnivorous mammals now known as mesonychians, but Philip Gingerich and colleagues recently argued that the newest paleontological evidence (see below) indicates that whales evolved directly from an ancestral artiodactyl. Within the Artiodactyla, recent immunological and DNA hybridization experiments specifically show a close relationship of whales to the hippopotami (**Fig. 5.1**). Regardless, one of the great mysteries of evolution and mammalian adaptation has been solved: Seagoing whales are descended from fully terrestrial mammals. We can piece the whole story together from fossil evidence that has become available only within the past 20 years.

Characteristics of Modern Whales

The Earth is apparently between 4.5 and 5 billion years (Gy) old. Life has been around, in one form or another, for at least 3.6 billion of those years. Vertebrate animals, those with backbones, have been on the planet for hundreds of millions of years. It seems utterly amazing, then, to realize that the largest animal that ever lived on Earth is alive today, at this single tick of the geological clock. Larger than *Ultrasaurus, Seismosaurus,* and every other dinosaur, the Blue Whale dwarfs them all, and we are fortunate enough to be contemporaries of this phenomenal creature. Large adults can approach 100 feet (about 30 meters) long and weigh in excess of 200 tons, heavier than a whole herd of elephants.

Whales, especially members of the Mysteceti like the Great Blue, are rather specialized animals. Their skulls are easily recognizable, with the dorsal blow holes, long attenuated snouts, and a skull specialized for underwater com-

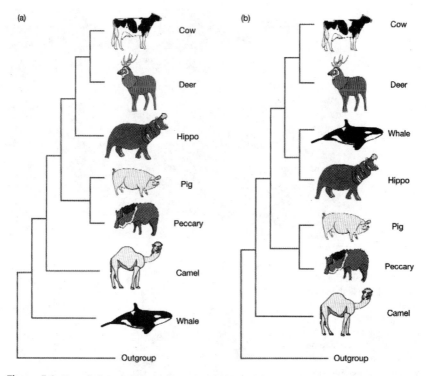

Figure 5.1 Two phylogenetic hypotheses for whales and their relatives. In (a) whales are considered to be related in a general way to other artiodactyls. In (b) they are specifically considered as a "sister" group to the hippos. Current evidence supports (b). The outgroup for comparison was the odd-toed ungulates, the Perissodactyla. (*Evolutionary Analysis,* 2e, by Freeman/Herron, copyright 2001. Reprinted by permission of Pearson Education, Inc., Upper Saddle River, NJ.)

munication. There are no living animals with which their skeletons can be confused. But there have always been hints of their ancestry, as a photograph of a female humpbacked whale attests (**Fig. 5.2**). Here, in a picture taken from a publication of the American Museum of Natural History in New York, Roy Chapman Andrews shows us the unmistakable evidence of tiny hindlimbs. As we have seen, it is evidence such as this that helps support the general theory of evolution. This observation leads us to the hypothesis that whales are descended from terrestrial animals with fully developed hindlimbs. Recently, the fossil record, our natural laboratory, has provided evidence confirming the predictions of this idea. However, before we examine this evidence, we need a brief primer of whale biology, relationships, and anatomy to better understand what we see in the fossil materials.

Whales are placental mammals, although their skin is basically naked. The females possess mammae and they suckle their young. Although whales are

(a)

(b)

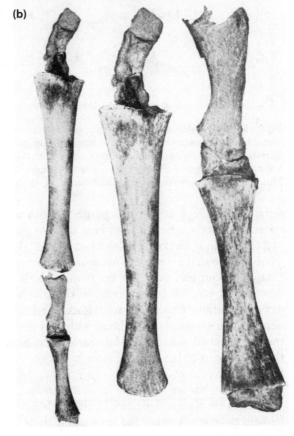

Figure 5.2 Vestigial external hind limbs in a humpback whale. In (a) is the rear end of the whale, with the limb in place. In (b) the elements have been dissected out for study. On the left is the entire limb from one side, in the center are the first two elements, identified as cartilaginous femur and bony tibia, and on the right are a cartilaginous tarsal element and bony metatarsal. ([a] Image #24617. Photo by R.C. Andrews/American Museum of Natural History Library. [b] Image #2A1158. Photo by R.C. Andrews/American Museum of Natural History Library.)

capable of extraordinarily deep dives, they must return to the surface to breathe. Unfortunately, every year some newborn whales die because they do not get to the surface in time for their first breath. The external nostrils of whales are located on the top of the head, the so-called blowhole. This position correlates with very reduced nasal bones. One of the observable evolutionary changes in the fossil record of whales is the reduction in size of the nasal bones and the posterior migration of the nasal openings on the snout. The skulls of cetaceans are rather bizarre structures when compared with those of other mammals (**Fig. 5.3**), and the cervical, or neck, vertebrae tend to be reduced and sometimes fused together. The front limbs are modified into flippers, and hindlimbs are vestigial or absent. The body posterior to the midline is designed to act with the enlarged tail fluke as a propulsive device in a porpoising, or up and down, motion, known as "caudal oscillation" in the scientific literature. The sacral vertebrae are separate to facilitate this movement. Interestingly, the sense of smell is likely absent in most whales, though the sense of taste is apparently well developed. Even more strange is the fact that whales do not possess vocal cords. The varied sounds they produce, both at extremely high and low frequencies, are generated by air moving between various sacs and the lungs in a closed system. Whales, like bats, are capable of using high frequency sound, or sonar, in a process called *echolocation,* to navigate in their surroundings. They use considerably lower frequency sounds, in the human audible range, for social communication.

Living cetaceans possess an anatomically unique bony auditory apparatus. As in all mammals, whales have three middle ear bones partially enclosed by the tympanic bone. In whales, the tympanic bone has a thickened medial ridge called an involucrum, and the involucrum further has an S-shaped bony fold known as the sigmoid process. Although their functions are unknown, this bony configuration is found only in whales, and consequently any fossil species that has it (if confirmed by other features) can be considered a whale. Furthermore, extant whales hear in an unusual way. Unlike terrestrial mammals, they do not use their external ear as the primary sound sensor. There is no external earlobe, or "pinna," in whales, and the ear opening is plugged. Terrestrial mammals such as ourselves, hearing in the less dense medium of air, transmit sound vibrations from an external membrane called the tympanum through the middle ear region. The pinna functions in focusing sound waves. In whales, sound vibrations are received by the lower jaw and are transmitted to the middle ear region by a special "fat pad" that fits in an enlarged mandibular foramen, a hole that covers much of the middle part of the inner side of the lower jaw. As we shall see, the earliest whales did not have this specialized system, though it apparently evolved fairly early in whale history.

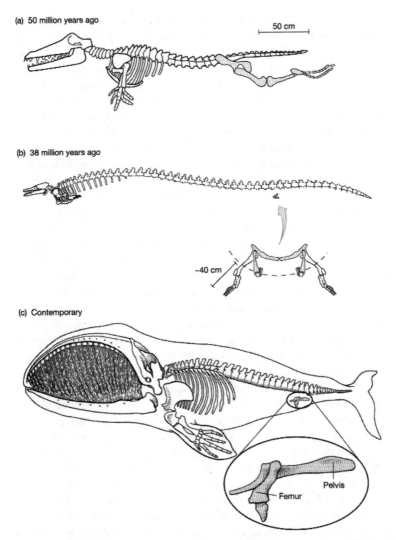

(a) 50 million years ago

50 cm

(b) 38 million years ago

~40 cm

(c) Contemporary

Pelvis

Femur

Figure 5.3 Transitional forms in whale evolution. *Ambulocetus natans* (a) was likely amphibious and had a full hind limb and pelvic girdle attached to the vertebral column. In *Basilosaurus isis* (b) the hind limbs were reduced and the pelvic girdle was free from the vertebrae. The hind limbs, rather than being used for swimming, may have functioned for grasping during copulation. In contemporary whales such as the blue whale, the hind limbs usually do not develop (but see Fig. 5.2) and the pelvic girdle is free from the vertebrae. These species do not represent a phyletic continuum; rather each is an example of a general evolutionary grade found in the fossil record. ([a] Reprinted with permission from J.G.M. Thewissen, S.T. Hussain, and M. Arif, 1994. Fossil evidence for the origin of aquatic locomotion in archaeocete whales. *Science* 263: 210–212. [b,c] Reprinted with permission from P.D. Gingerich, B.H. Smith, and E.L. Simons, 1990. Hind limbs of Eocene *Basilosaurus*: Evidence of feet in whales. *Science* 249: 154–159. Copyright 1990, 1994, American Association for the Advancement of Science.)

Hindlimb structures in whales are of particular significance to us here. Hindlimb buds are known from many living cetacean embryos, although they may be resorbed before birth. In some species, these vestigial structures are retained into adulthood, although the condition where the bones protrude from the body, as in the humpback whale noted in Figure 5.2a, is extremely rare. Even the most complete of these structures include only rod-like remains of the pelvic bones, femora, and sometimes the tibiae, usually embedded in muscle tissue. Foot bones are almost never developed, though Andrews interpreted one of the bones from the humpback as a metatarsal.

Whales are divided into two main groups, the toothed whales, or Odontoceti, and the baleen whales, or Mysteceti. Ten families are recognized, with the odontocetes being the most diverse. The familiar dolphins and porpoises are odontocetes, as is the huge sperm whale made famous by Herman Melville's *Moby Dick*. Curiously, no teeth are present in some members of one odontocete family, the Ziphiidae, but it is clear their ancestors possessed them, and intermediate conditions (a few teeth) are found in some members of the same family. Odontocetes feed on squid, fish, and bivalves, depending on the group. Modern odontocetes are functionally homodont, meaning that all the teeth have the same shape. This is a condition that seems to be associated with feeding on fish and crustaceans.

The largest species are mysticetes, such as the gray and blue whales. Some of these species are well adapted to life in the open oceans, the pelagic realm, and all demonstrate the peculiar epidermal plates in the roof of the mouth, called baleen, which are used to strain out zooplankton that mysticetes feed on. The word "zooplankton" means "floating animals" and includes many kinds of small crustaceans related to shrimp, called "krill," that occur in abundance in the deep oceans. Teeth are absent in mysticetes, though tooth buds are found in fetuses.

Early Whales: Transition from Land

About 48 Ma, in an inland bay of the ancestral Mediterranean Sea known as the Tethys seaway, strange creatures known as *Pakicetus inachus* cavorted and fed. *Pakicetus* is a member of an ancient family of whales, the Pakicetidae, recovered only from early Eocene deposits in Pakistan and India. The Pakicetidae is included in a larger group known as the Archaeocetes ("ancient whales"). Pakicetids are the oldest and least derived cetaceans and were relatively small as whales go, perhaps the size of a wolf. Although we know something about the skull and dentition of *Pakicetus* and can speculate about its feeding habits, its postcranial skeleton has unfortunately not been recovered. The eyes were located on the top of the skull, suggesting an amphibious life-style. The dentition of *Pakicetus* was partitioned into the recognizable units of land mammals, with distinct incisors (3/3), canines (1/1), premolars (4/4), and molars (3/3). Each pair refers to the number in the upper and lower dentition, respectively. Adding all together and multiplying by two gives the total number of teeth in the mouth. The number of teeth in *Pakicetus* is the underived or primitive condition for mammals. Deciduous teeth from juveniles as well as teeth from adults have been found. The arrangement and anatomy of the ear bones and posterior area of the skull (called the basicranium) in *Pakicetus* led its discoverers to suggest that it lacked both the ability to hear directionally in water and to dive deeply. In modern whales the ear region of the skull is specially designed for sound transmission under water and contains sinuses that can be engorged with blood to maintain pressure during deep dives. *Pakicetus* seems to have had a hearing apparatus more similar to that of terrestrial mammals, likely with a fully functional tympanic membrane. Recent geological evidence indicates that *Pakicetus* lived in freshwater streams, and according to P. Gingerich and D. Russell, faunal evidence from the Chorlakki locality in Pakistan suggests that *Pakicetus* spent a considerable time out of the water. For example, most of the other animals associated with *Pakicetus* were fully terrestrial, and the dominant invertebrates from the locality are land snails.

The earliest known whale, another pakicetid named *Hilmalayacetus,* from the Subathu Formation of northern India, was recovered from marine strata associated with a marine fauna. *Himalayacetus,* at about 53.5 Ma, is known only from a left mandible with the last two molars. Described by S. Bajpal and P. Gingerich in 1998, the jaw has none of the advanced specializations for hearing seen in extant cetaceans.

Another remarkable whale named *Ambulocetus natans* was found 120 meters above the remains of *Pakicetus* (**Fig. 5.3**). *Ambulocetus* is the only member of the family Ambulocetidae and is structurally more advanced than the pakicetids. The age of *A. natans* is considered to be approximately 47 to 48 Ma. According to J. G. M. Thewissen and colleagues, *Ambulocetus* was about the size of a male sea lion, maybe 300 kg (thus considerably larger than either *Himalayacetus* and *Pakicetus*). It possessed the long snout and expanded braincase characteristic of whales but also demonstrated the hindquarters of an animal that may very well have spent a considerable amount of time on land. This can be seen easily by examining the illustrations in Figure 5.3. The whale involucrum was present, but the region of the sigmoid process was not preserved. In comparison with a modern whale, the skeleton of *Ambulocetus* had distinct similarities to that of a terrestrial animal. The neck was long and mobile, and the hindlimbs were substantial. The tail vertebrae indicate that the tail did not possess a fluke. Propulsion was provided by up and down porpoise-like motions of the body as in modern whales, coupled with pulses from the hindfeet, as in modern sea otters. However, unlike these otters, it was the feet, not the tail, that provided the power stroke. Amazingly, the toes of *Ambulocetus* ended in tiny hooves, much like the artiodactyls mentioned above. The forelimb of *Ambulocetus* was rather flexible. The hand was long and possessed five digits, the primitive number for terrestrial vertebrates. Nevertheless, the hands almost certainly provided only navigational guidance rather than aiding in locomotion in the water. On land, the forelimbs were splayed out, as in some seals, providing support for the body and balance during motion. As indicated by the associated animal and plant remains (e.g., marine snails, oysters, etc.), *Ambulocetus* inhabited shallow marine waters in the ancient Tethys Sea. Although the hindlimb structure of *Pakicetus* is unknown, that of *Ambulocetus,* as noted above, clearly indicates the animal could support itself on land. Reproduction and development may have taken place entirely out of water in these primordial cetaceans. Taken in total, the evidence suggests that ambulocetids filled a niche similar to that of modern crocodiles.

Another cetacean family, distributed in near-shore deposits of Pakistan and western India between about 43 and 49 Ma, is the somewhat aberrant Remingtonocetidae, named after an American paleontologist who worked with fossil whales early in the twentieth century. They had long snouts and small eyes located on the side of the skull. Their middle ear region was very large (like modern odontocetes) and they had large powerful tails. Their legs were relatively short and definitely could support the body on land. According to J. G. M. Thewissen and S. Bajpal, they would have appeared similar to long-snouted crocodiles. The different genera varied in size from perhaps 40 to 300 kg.

Whales Take to the Sea

New discoveries of fossil whales from the Balochistan Province of Pakistan give us a more accurate picture of the anatomy and likely habits of early whales as they became more attuned to shallow marine environments. In 2001, Philip Gingerich and colleagues described the skull of a new genus and species called *Artiocetus clavis* and the skull and almost complete skeleton of a new species of *Rhodocetus, R. balochistanensis.* A partial skeleton of another protocetid, *Rhodocetus kasrani,* had been described a few years earlier, but the hindlimbs were unknown. These whales, members of the family Protocetidae, died and were buried in Tethyian sediments about 47 Ma. There are obvious similarities in overall body form between *Rhodocetus* and the earlier *Ambulocetus,* but in *Rhodocetus* the hands and feet are distinctly longer. Both animals were probably otter-like pelvic paddlers, and the feet may have been webbed. The skeleton of *Rhodocetus* shows distinct similarities to that of terrestrial mammals but is also obviously specialized for locomotion in water. Of special importance is the relatively large pelvic region, which Gingerich and colleagues interpret to indicate that the animal could support its body on land, like modern seals and sea lions. Although the femur of *Rhodocetus* is only about 60% to 70% as long as it is in the earlier archaeocete *A. natans*, the sacrum still retains a connection with the pelvis, unlike the condition in later Eocene *Basilosaurus, Durodon,* and modern whales, in which this connection is lost (**Fig. 5.4**). Although protocetids probably spent some time on land, their hindlimb anatomy was reduced to the point that terrestrial movement would not have been easy. *Rhodocetus* displays some further advanced, or derived, skeletal features akin with later whales, such as shortened cervical (neck) vertebrae and unfused sacral vertebrae in addition to the reduced femur. The shortened neck region provides rigidity to the anterior body, and the other features help to streamline the posterior body trunk. Finally, the form of the vertebrae, with high spines, indicates that powerful muscles responsible for dorsoventral flexion of the body were present and probably provided the dominant propulsive force, likely through a tail fluke, though this terminal caudal area has not been found in *Rhodocetus*. The eyes in *Rhodocetus* were on the side of the skull, and the nasal openings would have been somewhat posterior to the end of the snout. Check out the painting of a hypothetical *Rhodocetus balochistanensis* on the cover of the September 21, 2001 issue of *Science* magazine. Behold another "missing link."

The next stage in whale evolution is represented by the Families Basilosauridae and Dorudontidae. Whales of these families have been recovered from deposits in India, Pakistan, New Zealand, Africa, and North America. *Basilosaurus isis* (along with some dorudontids) was found in the Fayum beds of north central Egypt, of middle Eocene age, about 37 Ma. The name *Basilosaurus* may seem strange for a whale, but that was because the individual who named the first fossils of this animal, R. Harlan, thought it was a reptile, as this quote from Melville (1952, p. 422) attests:

> *... by far the most wonderful of all cetacean relics was the almost complete vast skeleton of an extinct monster, found in the year 1842, on the plantation of Judge Creagh, in Alabama. The awestricken credulous slaves in the vicinity took it for the bones of one of the fallen angels. The Alabama doctors declared it a huge reptile, and bestowed upon it the name* Basilosaurus. *But some specimen bones of it being taken across the sea to Owen, the English anatomist, it turned out that this alleged reptile was a whale, though of a departed species....*

Work at the Fayum beds has unearthed a treasure trove of early whales: 243 skeletons of *Basilosaurus* and 77 partial skeletons of dorudontid archaeocetes. In basilosaurids and dorudontids we can see a condition inter-

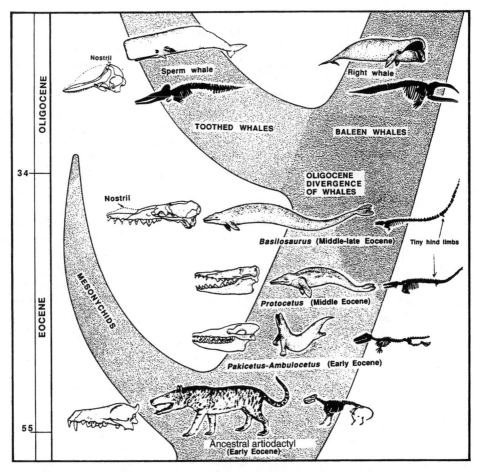

Figure 5.4 A possible phylogeny for whales. (D.R. Prothero, 1994. Mammalian evolution. In D.R. Prothero, and R.M Shroch, eds. *Major Features of Vertebrate Evolution.* Pittsburgh: Paleontological Society.)

mediate between *Ambulocetus* and later whales. The skeletons of these animals are instantly recognizable as those of whales. In both taxa, the external nares were farther back on the rostrum than in *Rhodocetus* (a trend that continues into modern whales). A sacrum was absent, and the pelvis was much reduced in size from the condition in *Ambulocetus*. The pelvis would not have been able to support the body of these animals on land. The first digit of the foot was absent in *Basilosaurus* and the second reduced to a tiny nubbin. Other foot bones were present, unlike the condition in modern whales, where in the vestigial hindlimb structures usually no foot bones are retained. The foot bones of *Basilosaurus* show distinct similarities to those of terrestrial ancestral artiodactyls. It seems likely that the femur was contained entirely within the body wall, and this observation, combined with other considerations, led P. Gingerich, B. H. Smith, and E. Simons to suggest that the hindlimbs of *Basilosaurus* were too small to assist in swimming or to support the body on land. They interpreted the hindlimbs of *Basilosaurus* as copulatory guides such as are found on the pelvic fins of modern sharks, facilitating sperm transfer in an aquatic environment. *Basilosaurus* itself was probably not on the direct line to modern species because it had a large (ca. 20 meter) and rather elongate body shape. Dorudontids were smaller, about the size of living dolphins.

A new dorudontid cetacean from middle-to-late Eocene sediments in South Carolina, described by Mark Uhen and Philip Gingerich in 2001, may better form the bond between ancient and modern whales. The holotype specimen of *Chrysocetus healyorum* includes much of the skeleton of a juvenile individual, although the hindlimbs are not preserved. The pelves are at a comparable stage of evolutionary reduction to *Basilosaurus* and could not have functioned on land. Perhaps the most unusual feature of *Chrysocetus,* and the one that sets it apart from other archaeocetes and on the line toward modern whales, is its presumed mode of dental development. As noted by Uhen and Gingerich (2001, p. 20):

> Dental development of C. healyorum *did not follow the same trajectory as other archaeocetes. It either replaced the deciduous dentition very early in development relative to skeletal maturation, or it never had deciduous teeth and instead erupted teeth of adult morphology early in development relative to skeletal maturation. All other archaeocetes retain the pattern of replacement of the deciduous dentition with an adult dentition (diphyodonty).... In contrast, modern suborders of Cetacea ... both erupt only a single set of teeth (monophyodonty). No living or fossil odontocetes replace their teeth..., and mysticetes develop a single set of teeth that is later resorbed in utero....*

Chrysocetus then makes a good generalized common ancestor for both the Odontocetes (retaining teeth as adults) and the toothless mysticetes. And the story does not end there. It is impossible in this treatment to review the characteristics of every fossil whale, but a number of toothed mysticetes are known, including *Llanocetus* from the early Oligocene of Seymour Island, in the Antarctic region, and *Aetiocetus,* from the late Oligocene of the northeastern Pacific rim. Sperm whales (Physeteridae) currently hold the record for being the most ancient of the modern whales. Primitive sperm whales, such as *Diadophorocetus* from the western south Atlantic region, had functional upper and lower teeth, whereas more advanced species, such as the living sperm whale *Physeter catodon* (the only living sperm whale), have reduced upper teeth. All but a few modern whale families had appeared in the fossil record by Miocene time, so it seems pretty clear that they diversified rapidly from their Oligocene ancestors.

In summary, let us list the evidence that conclusively demonstrates the evolution of whales from ancient terrestrial artiodactyls:

1. Many modern whales possess vestigial hindlimb structures.

2. Many modern whales display hindlimb buds during the fetal stage, which are lost as the animal develops further.

3. Modern mysticete whales, which do not possess teeth as adults, display tooth buds during development. Ancient adult mysticete whales with teeth are known.

4. All ancient whales, the Archaeocetes, possessed teeth.

5. Ancient archaeocetes possessed a dental formula more similar to ancestral terrestrial mammals than to modern whales, but the form of these teeth was intermediate between the two groups.

6. The ear and lower jaw region of ancient whales was more similar to that of terrestrial mammals than to fully aquatic modern whales. Intermediates between the two are known.

7. The hindlimbs of *Ambulocetus,* an ancient whale, are intermediate between terrestrial mammals and modern whales and were used in the power stroke to propel the body.

8. The ends of the digits of *Ambulocetus* and *Rhodocetus* display tiny hooves. These animals obviously spent part of their lives on land but had definite aquatic adaptations as well.

9. Dorudontid and basilosaurid whales occur in sediments younger than the more primitive archaeocetes and are structurally intermediate between archaeocetes and modern whales.

10. The modern sperm whale lacks functional upper teeth. Ancient physterid whales with functional upper and lower teeth are known.

Suggested Readings

Andrews, R. C. 1921. A remarkable case of external hind limbs in a humpback whale. *American Museum Novitates* No. 9:1–6.

Bajpai, S. and P. D. Gingerich. 1998. A new Eocene archaeocete (Mammalia, Cetacea) from India and the time of origin of whales. *Proceedings of the National Academy of Sciences* 95:15464–15468.

Berta, A. 1994. What is a whale? *Science* 263:180–181.

Eisenberg, J. F. 1981. *The Mammalian Radiations.* University of Chicago Press, Chicago.

Fordyce, R. E. 1980. Whale evolution and Oligocene southern ocean environments. *Palaeogeography, Palaeoclimatology and Palaeoecology* 31:319–336.

Fordyce, R. E. and L. G. Barnes. 1994. The evolutionary history of whales and dolphins. *Annual Review Earth & Planetary Sciences* 22:419–455.

Gingerich, P. D. 1994. The whales of Tethys. *Natural History* 103:86–88.

Gingerich, P. D. and D. E. Russell. 1990. Dentition of early Eocene *Pakicetus* (Mammalia, Cetacea). *Contributions to the Museum of Paleontology*, University of Michigan 28:1–20.

Gingerich, P. D., B. H. Smith, and E. L. Simons. 1990. Hind limbs of Eocene *Basilosaurus*: evidence of feet in whales. *Science* 249:154–157.

Gingerich, P. D., M. ul Haq, I. S. Zalmout, I. H. Khan, and M. S. Malkani. 2001. Origin of whales from early artiodactyls: hands and feet of Eocene Protocetidae from Pakistan. *Science* 293:2157–2336.

Gingerich, P. D., N. A. Wells, D. E. Russell, and S. M. I. Shah. 1983. Origins of whales in epicontinental remnant seas: new evidence from the early Eocene of Pakistan. *Science* 220:403–406.

Gould, S. J. 1994. Hooking leviathan by its past. *Natural History* 103:8–15.

Prothero, D. R. 1994. Mammalian evolution; pp. 238–270. In D. R. Prothero and R. M. Schoch (eds.), *Major Features of Vertebrate Evolution, Short Courses in Paleontology* 7. The Paleontological Society, University of Tennessee, Knoxville.

Thewissen, J. G. M. and S. Bajpal. 2001. Whale origins as a poster child for macroevolution. *Bioscience* 51:1037–1049.

Thewissen, J. G. M., S. T. Hussain, and M. Arif. 1994. Fossil evidence for the origin of aquatic locomotion in archaeocete whales. *Science* 263:210–212.

Uhen, M. D. and P. D. Gingerich. 2001. New genus of durodontine archaeocete (Cetacea) from the middle-to-late Eocene of South Carolina. *Marine Mammal Science* 17:1–34.

6

The Kentucky Derby in Deep Time: Evolution of Horses During the Cenozoic

"It is probably anthropomorphic to speak of 'progress' in evolution; this implies a purposeful directionality which seems to be lacking."

When we think of horses, we think of their power, their speed, and their grace. We consider them as domesticated animals, but their history in human animal husbandry actually only goes back to perhaps what we refer to as the late Neolithic Period in human cultural evolution, at the end of the fourth or beginning of the fifth millennium BC. Before that time our ancestors probably viewed them, along with all other large mammalian herbivores, as potential food. In grade school we learn that horses were introduced into the New World by the Spanish conquistadors in the sixteenth century AD, but less well known is the fact that much of horse evolution took place in the New World over a period of almost 60 million years. Unfortunately, the last native horses became extinct in North America about 10,000 years ago. The evolution of horses has long been a subject for textbooks because so much fossil material was made available early in the twentieth century, mostly from the efforts of paleontologists and collectors at the American Museum of

Natural History in New York. Changes in foot structure, skull morphology, and body size of horses provided powerful and incontrovertible proof of the theory of evolution as proposed by Darwin, and it is not surprising that the general and scientific literature often used the horse as classic proof for this theory. There has been some controversy as to how the horse story was told, particularly the extent to which early authors were influenced by the concept of "phyletic gradualism" (see Chapter 2), but the evidence has always been overwhelming. I review some of it here. Much of this information is taken from the modern work of my colleagues, Bruce MacFadden, Donald Prothero, and Richard Hulbert. MacFadden's book, *Fossil Horses: Systematics, Paleobiology and Evolution of the Family Equidae*, represents the most complete history of any mammal group in the fossil record and is highly recommended for the amateur and specialist alike.

What Are Horses?

First, horses are part of a general grouping known colloquially as "ungulates," the hoofed mammals. They make up about one third of the living and extinct mammals at the generic level. Ungulates are an extraordinarily diverse lot, including the modern camels, pigs, horses, rhinos, tapirs, sea cows, hippos, elephants, deer, antelopes, buffalos, and cows, and even whales, because it is clear that whales are descended from ancient hoofed ancestors. Horses are grouped together in the Order Perissodactyla with other ungulates that characteristically have "odd toes," meaning they have either one or three toes on their feet. Some, like the primitive perissodactyls and tapirs, have four toes on the forefeet, but all are allied in that the primary weight-bearing digit is the central one of the hand and foot. Often, the two lateral digits on either side are retained, though they may not be highly functional. The perissodactyls include three major groups: the Hippomorpha, the horses and extinct palaeotheres; the extinct Titanotheriomorpha, or brontotheres; and the Moropomorpha, the modern tapirs, rhinos, and their extinct relatives. An enigmatic group is the hyraxes, African animals that look like large rodents. Currently, hyraxes are viewed as the sister group of perissodactyls, and both are likely descended from a generalized ancestor like *Radinskya* of Paleocene age in China.

The horses, or Family Equidae, are allied by a series of structural characteristics that are not discussed here in detail. Suffice it to say features of the cranium, including the absence, presence, and position of certain blood vessel and nerve conducting holes (foramina, the plural of "foramen"), and the molars set them apart from other perissodactyls. As with all animal clades, as one examines ancestral fossils back further in time, they all seem to converge structurally on one another, and it is at this point that true evolutionary relationships can sometimes get confused (**Fig.6.1**). For example, the earliest recognized horse for decades was considered to be "eohippus," the name of which was later changed to *Hyracotherium.* Most of these early Eocene species are quite similar, but modern phylogenetic analyses are beginning to tease them apart into their respective clades. After this point in time it becomes easier to spot a "horse," and this account will deal primarily with these later animals.

Modern equids include a single genus, *Equus,* with perhaps as many as eight species, informally grouped as the horses, zebras, and asses. They are all similar in body shape, with long necks and heads, one functional digit on each limb, long tails, and two mammae. Living equids literally walk on the tips of their middle digits, which are widened and rounded. The radius and ulna bones of the forefeet are fused and the radius is very reduced, so the weight is born entirely by the ulna. Likewise in the rear legs, the fibula is reduced and fused to the tibia. The skull and teeth are designed for grazing, and the premolars are "molariform" (like the molars), creating a long wide grinding surface to process large quantities of fodder. Zebrines (zebras), caballines (horses), and hemionines (asses) can generally be separated on the basis of characters of the lower molars.

Two wild horses are believed to have existed in Europe until fairly recently, the tarpan (*Equus caballus*) and Przewalskii's horse (*E. przewalskii*), although some recent studies consider the tarpan a domesticated version of Przewalskii's horse. Modern domesticated horses (including those reintroduced into North America) are likely descended from one or both of these taxa. Zebras (*E. zebra, E. burchelli, E. grevyi, E. quagga*) are now limited to Africa, and the modern asses and onagers (*E. asinus, E. hemionus, E. kiang, E. onager*) were historically distributed throughout southern Asia through the Middle East into northern Africa. Prehistorically, the picture is quite different, because members of these groups were also distributed widely in North America through the late Cenozoic. A mule is the sterile offspring of a female horse and a male ass. Mules were bred for their endurance and surefootedness.

Orthogenesis and the Bushy History of Horses

Before Darwin, there was a prevailing philosophical and religious view, dating back primarily to the classic Greeks, especially Aristotle and Plato, that life was arranged in an immutable scheme with the least complex organisms "below" the most complex and with humans as the pinnacle of God's grace. Aristotle's "ladder of perfection" is a reflection of the notion. Some evolutionary biologists today believe that many, if not most, of the scientists of the late nineteenth and early twentieth centuries were influenced by this line of thinking and that they interpreted Darwin's *Origin* as the story mostly of

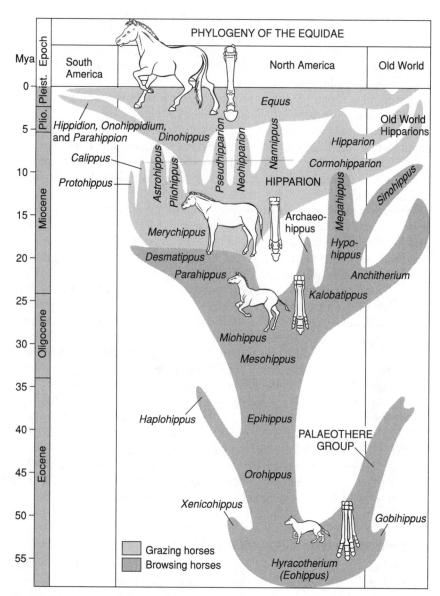

Figure 6.1 The evolution of horses can be best characterized as a bushy tree, with many experiments in size and anatomy going on simultaneously. Over millions of years, selection favored larger horses able to run fast, undoubtedly a response to the world's emerging grasslands and a foraging shift from browsing to grazing. (Adapted with additions from B.J. MacFadden, 1992. *Fossil Horses: Systematics, Paleobiology, and Evolution of the Family Equidae.* Cambridge, UK: Cambridge University Press.)

species changing gradually and monotonically through millions and millions of years, leading to modern horses as more "perfect" than earlier horses. This interpretation, with its inherent acceptance of progress, is known as "orthogenesis."

We now realize that the evolution of horses was characterized by a complex radiation of species during many time periods and that both "primitive" and "advanced" horses existed simultaneously, at least for brief periods of time. Nevertheless, although our intellectual ancestors were a bit primitive in their taxonomy, they were not the idiots many modern scholars make them out to be. George Gaylord Simpson, for example, knew very well that there was a complex adaptive explosion of many kinds of horses during the Cenozoic. Yet, the earliest horses were all small, later horses included both small and large species, and the latest species, those that existed during the Pleistocene, were all large. Modern horses average about 500 kg, with the Clydesdales topping out over 1000 kg. Additionally, there has been a general trend toward reduction in the number of toe bones and fusion of skeletal elements (e.g., radius with ulna, fibula with tibia) as this large size progression unfolded. Horse evolution does present, as Harold Blum would say, evidence of "time's arrow" or the stamp of irreversibility often referred to as "Dollo's Law." Horses did not start out large and get smaller on average through time (though dwarfing lineages and clades are known) nor did their skeletal elements ever "unfuse." Does this represent "progress" in any way, or is there a possibility that evolutionary trends in horses represent, as Gould has suggested (see Chapter 2), another example of the Drunkard's meaningless walk? It is probably anthropomorphic to speak of progress in evolution; this implies a purposeful directionality that seems to be lacking, but it may be that later grazing horses are better at what they do than early ones: better at running from predators, more efficient herbivores at processing grass, and so on. But that kind of a comparison would only be appropriate if the earlier species were living in the environments of the later ones. History seems to say to us that those species alive at any point in time were manifestly well adapted for their environments; it is just that the environments (including other organisms) never remain stable. As the Red Queen hypothesis suggests (see Chapter 3), animals do the best they can, which is never quite good enough, as everything around them changes; natural selection played out over long periods of time.

Earliest Horses

Eocene horses were distributed in a Holarctic fashion, meaning at temperate and northern latitudes throughout the world: in North America, Europe, Canada, and Asia. This kind of simultaneous distribution in the early Eocene presents problems in determining the center of origin for horses, but the presence of *Radinskya,* considered a possible ancestral form, in late Paleocene beds of China suggests a southeast Asian epicenter of evolution and dispersal.

These dawn horses were small, less than 5 kg in size, and possessed teeth that were small, low-crowned (brachydont), with discernible cusps and lacking cement. They were thus best designed to browse low vegetation such as twigs, stems, and leaves, much as deer do today. *Hyracotherium* possessed four digits on the forefeet and three on the hindlimbs (abbreviated as 4/3), but even then the primary weight-bearing digit was the central one. These early horses are collectively known as the "hyracotheres" and include such taxa as *Protorohippus, Hyracotherium, Pachynolophus, Anchilophus,* and *Cymbilophus.*

During the middle Eocene, around 50 million years ago (Ma), the dispersal corridor for equids between the Old and New Worlds mostly closed, presumably because of continued seafloor spreading in the Atlantic Ocean. Before that time there had likely been a connection from North America across Greenland and into northern Europe. This route is sometimes referred to as the "DeGeer" dispersal route. The Old World horse taxa that subsequently became isolated are known as the Palaeotheriidae, becoming extinct in the Old World during the Oligocene. The Equidae proper evolved in North America from those Eocene taxa remaining there.

The Horses of Middle Time: Allometry of Skull Shape

Browsing horses continued to dominate in North America for more than 30 million years, through the Oligocene and into the early Miocene epochs. Their molars, although much larger than their Eocene ancestors, were still relatively low crowned and cuspate. Three toes were present on the fore- and hindfeet, and the lateral digits touched the ground. But the dental patterns that were to herald those of the later grazers were apparent in such genera as *Mesohippus* and *Miohippus.* Additionally, the skulls of Miocene browsers began to evolve toward those of modern horses.

Allometry refers to the differential growth rates of various parts of the body during development or evolution. In Eocene horses such as *Hyracotherium,* the front part of the skull was relatively short as compared with the rear part. Later horses show an *allometric* response of skull growth relative to body size increase (**Fig. 6.2**); the front part of the face grows faster in later horses, resulting in a longer muzzle. This change in the speed of developmental timing progressed to even greater extent in *Equus,* so that the rear part of the skull makes up a much smaller proportion of the overall skull than it did in the hyracotheres. The functional significance of this appears to relate to foraging and food processing. Through time there was positive selection for those horses that were more efficient at metabolizing food and escaping predators. Those with larger molariform premolars could pulverize more food and were also those in which the

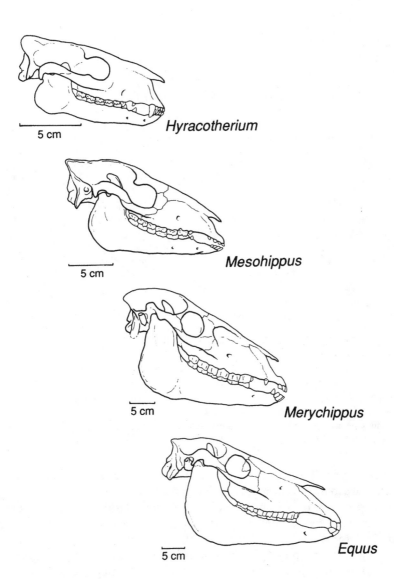

Figure 6.2 Changes in cranial proportions of horses through time. Note the relative elongation of the muzzle anterior to the orbit (eye socket). (From B.J. McFadden, 1992. *Fossil Horses.* Cambridge, UK: Cambridge University Press. p. 199.)

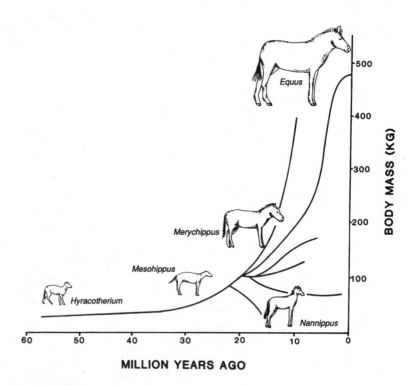

Figure 6.3 Body size evolution in horses over the past 50 million years. Although there have been small horses during most time periods, there has also been a general overall increase in size through time, initiated during the development of the world's great grasslands. (From B.J. McFadden, 1992. *Fossil Horses*. Cambridge, UK: Cambridge University Press. p. 221.)

muzzle was longer. These horses also tended to be larger and faster, two important attributes in the predator/prey evolutionary "race."

Mesohippus was the first horse with only three functional toes on the forefoot, as opposed to the four of hyracotheres. The side toes possessed pads, but the central digit and pad were much enlarged. The radius/ulna and tibia/fibula in this taxon displayed various levels of fusion. A truly intermediate condition between the hyracotheres and later grazing equids.

During the early Miocene, about 24 Ma, equids again dispersed to the Old World, across the most commonly used dispersal route through the later Cenozoic, the Bering Straits. As sea level lowered at various times, a connection was established from Siberia across the Aleutian Islands to Alaska. This area was so large that it is sometimes considered the "subcontinent" of Beringia. This dispersal event was limited exclusively to the large browsing anchitheres, such as *Kalobatippus* and *Anchitherium*.

Emergence of Grasslands and the Origin of Modern Horses

The great grasslands of the world are among the most productive of the planet's ecosystems. The pampas of Argentina, the Great Plains of North America, the grassy steppes of Eurasia, and the savannas of Africa seem to have developed from Oligocene through middle Miocene time, and this change in plant distributional patterns was to have a profound effect on mammalian evolutionary patterns. Savannas especially were even more widespread than they are today. It is during this time that we see a pronounced radiation of equids more adapted to grazing than their predecessors, characterized especially well by a generalized ancestral genus known as *Merychippus*. *Merychippus* was morphologically advanced over *Miohippus* in a number of ways. In the forearm, the ulna was reduced and partly fused with the radius. The central digit was much enlarged relative to digits II and IV, though both probably still touched the ground to prevent overextension of the forelimb. The enlargement of the central metapodial digit led in *Merychippus* to the first of the truly *unguligrade* rather than *digitigrade* horses. That is, *Merychippus* was the first horse to truly run on the tips of its fingers and toes. As a consequence of the extreme elongation of its metapodials, the wrist and ankle joints were elevated off the ground. Species of the genus *Merychippus* were also larger than those of *Miohippus,* averaging about 150 to 200 kg. Modern horses are about 500 kg (**Fig. 6.3**).

Merychippus appears to have evolved from *Parahippus* of the early Miocene, and it may have been a key dental innovation of the former that led to the success, not only of *Merychippus,* but of most later grazing equids. *Parahippus* was the first of the *hypsodont* horses, meaning that its molars were very high crowned compared with earlier horses. High crowns, composed of enamel (the hard mineral hydroxy apatite), were advantageous to grazing animals that ingested sand and other hard gritty materials with their food, as it in effect extended their lives. This innovation, which also appeared in other mammalian groups as grasslands appeared and spread, led to an explosion of many kinds of grazing mammals during the middle Miocene. Bruce MacFadden estimates that between 55% and 87% of mammalian communities were composed of grazing mammals of one sort or another.

At the end of the Miocene, about 5 to 7 Ma, many lush savanna environments gave way to arid steppe conditions, resulting in a significant decline in equid diversity, although some taxa persisted in southern refugia such as Florida for a while. Subsequently, about 4.5 Ma, *Equus* first appears, apparently derived from the Miocene *Dinohippus*. *Equus* and other late Cenozoic equids first demonstrate true monodactyly, the condition in which only a single digit in each foot touches the ground. Interestingly, Michael Voorhies reported a population of *Dinohippus* from the late Miocene Ashfall Fossil Beds in Nebraska in which some individuals are tridactyl and others monodactyl.

Other hypsodont horses persisted for a while, including the dwarf forms *Nannippus* and *Neohipparion,* but only Equus survived into and through the Quaternary Period (last 1.8 million years). In *Equus* we see the culmination of 58 million years of experimentation; the radius/ulna and tibia/fibula are completely fused. In the foot, the interosseus tendon almost completely replaces the muscle by the same name, resulting in the "springing" locomotion of modern horses. This structural modification results in a more powerful gait through the storage of potential energy in elastic ligaments and tendons. The well-known "stay" apparatus of modern horses, an anatomical modification allowing for the limbs to lock into position when horses are at rest, is first seen in a rudimentary fashion in *Dinohippus.* Modern horses also include the largest animals among the known members of the Equidae. Equid evolution nicely exemplifies "Cope's Rule," the tendency among many clades to evolve toward large size, though we now know not to interpret this to imply orthogenesis (or "directed" evolution). We also know, from the work of many authors, that small horse species appeared from time to time in the fossil record, including some like *Nannippus* that had hypsodont molars. There was no single anagenetic lineage leading from a single ancient hyracothere straight through to a modern species of *Equus.* Rather, there was considerable speciation and constant selection for those horse species best fit for their environment and a lot of experimentation going on, with many different horse species alive at any given time.

Suggested Readings

Azzaroli, A. 1990. The genus *Equus* in Europe; pp. 339-356. In E. H. Lindsay, V. Fahlbusch, and P. Mein (eds.), *European Neogene Mammal Chronology. Plenum Press,* New York.

Dalquest, W. W. 1988. *Astrohippus* and the origin of Blancan and Pleistocene horses. *Occasional Papers, Museum of Texas Technological University* 116:1–23.

Devillers, C., J. Mahe, D. Ambrose, R. Bauchot, and E. Chatelain. 1984. Allometric studies on the skull of living and fossil Equidae (Mammalian: Preissodactyla). *Journal of Vertebrate Paleontology* 4:471–480.

Edinger, T. 1948. Evolution of the horse brain. *Memoirs Geological Society of America* 25:1–177.

Eisenmann, V. 1986. Comparative osteology of modern and fossil horses, half-asses, and asses; pp. 67–116. In R. H. Meadow and H.-P. Uerpman (eds.), *Equids in the Ancient World.* Verlag, Wiesbaden.

MacFadden, B. J. 1992. *Fossil Horses: Systematics, Paleobiology and Evolution of the Family Equidae.* Cambridge University Press, New York.

Prothero, D. R. and R. M. Schoch (eds.). 1989. *The Evolution of Perissodactyls.* Clarendon Press, Oxford.

7

The Thanksgiving Day Dinosaur

"Within the animal kingdom there are hardly two classes which, at a glance, differ more from one another than the reptiles and the birds. Take, for instance, a nightingale and a tortoise, or the secretary bird warding off the poisonous bite of the African cobra: there seems to be no similarity between thems....

"Every move of the bird is characterized by its warm pulsating blood; its passions are strong, and its feeling so intense as to vent itself in song.

"The reptile seems to be the very opposite to all this. Sluggish and slow it creeps along; it takes sunshine and heat to stimulate it to action; the cold paralyzes its every movement.

"All these, however, are only superficial differences."
—*Gerhard Heilmann*

My colleagues John Ostrom, Jacques Gauthier, Luis Chiappe, Kevin Padian, and Bob Bakker are convinced that dinosaurs are still among us. We call them birds, but there are many characters of the skeleton that seem to ally modern birds with coelurosaurian dinosaurs. Depending on how the formal classification falls out, we may still be able to call birds birds, but they are also often included these days in more encompassing taxa that additionally incorporate such killing machines as *Deinonychus* and *Velociraptor*. Despite observed similarities, another paleontological camp, including Alan Feduccia and Larry Martin, suggests instead that birds and dinosaurs share a common ancestry from an as yet undiscovered extinct basal thecodont (generalized archosaur with roots of teeth in sockets). They claim that the similarities between dinosaurs and birds represent the results of evolutionary convergence; "homoplasy" in cladistic jargon. Convergent evolution occurs when two groups of animals come to resemble each other, not because of immediate close relationship, but because they are solving similar ecological problems, which requires similar anatomical solutions. In the case of birds and dinosaurs, they are additionally fairly closely related, so we would expect in any case to see many similarities. But to conclude that birds are descended from dinosaurs, investigators must determine that birds and dinosaurs share uniquely derived features (the "synapomorphies" of the cladist; see below) not shared with other groups. Therein lies the problem. It is always the choice of characters and the perception of the investigator that determines the classification, and Feduccia and Martin do not agree with the others on either their choice of characters or the interpretation of these characters. Until relatively recently we had only the astounding *Archaeopteryx* (**Fig. 7.1**) on which to base our conclusions, but discoveries in the past 10 years, especially from China, have provided a lot of new material that has yet to be exhaustively compared. We now have many toothy birds, some with long tails, some with short, and it turns out that feathers did not only evolve in birds. Paleontologists have recovered dinosaurs with feathery structures, and even *Velociraptor,* that agile carnivore made famous in Steven Speilberg's *Jurassic Park,* has occasionally been reconstructed with feathers (though there is no direct evidence for that).

The literature on bird origins and relationships provides a wonderful example of how science works at its best. Some of the most colorful characters in paleontology have contributed (and are still contributing) to this literature, and because of this interest avian evolution continues to be one of the most vibrant areas of paleontological inquiry. Not too long after the first dinosaurs were discovered, Thomas Henry Huxley, one of Charles Darwin's great admirers and acolytes and a renowned scientist and orator, proposed that dinosaurs were closely related to birds. However, other scientists of his time (like some in our time) considered the shared bipedalism of dinosaurs and birds to be the result of convergence and not true relationship, and the idea laid fallow for quite literally 100 years until it was revived by John Ostrom of Yale University in the late 1960s. In a series of carefully documented papers dealing with new fossil material of the Jurassic reptilian carnivore *Deinonychus* and other coelurosaurians, Ostrom crafted an evolutionary scenario that forms the basis of most of today's work. A number of investigators have since evaluated dinosaur and avian features with modern cladistic methods and essentially confirmed that birds are allied closely with coelurosarian theropod dinosaurs. By the principles of modern phylogenetic analysis, birds could be considered dinosaurs and therefore reptiles. As some of my colleagues are fond of telling me, although the Cenozoic Era is called "The Age of Mammals," there are far more living birds *and other reptiles* than there are living mammals. But we are getting ahead of the story. To demonstrate that *Archaeopteryx* is a link between

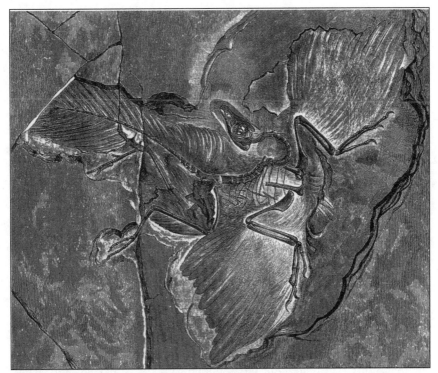

Figure 7.1 The Berlin specimen of *Archaeopteryx,* perhaps the most famous fossil specimen in the world. Note the teeth, clawed fingers, and long reptilian bony tail.

reptiles and birds, we should first examine some of the features found in living representatives of these two groups that help to establish a connection even in the absence of the fossil record.

Characteristics of Modern Birds

There are probably hundreds of detailed characteristics that could be listed indicating that birds and reptiles share a common ancestry, from which the following were drawn, many from Gerhard Heilmann's classic 1926 treatise, *The Origin of Birds*:

1. Similarity of form and position of eye, ear, and braincase (eye, for example, with a bony sclerotic ring);
2. Many similar bones in skull and skeleton;
3. Homology of feathers and scales;
4. Homology of beak of birds and horny anterior head plates of reptiles;

5. Digital claws in some adult birds;

6. Egg tooth in both groups;

7. A special flexor muscle, the *ambiens*, found only in birds and reptiles.

Additionally, birds and reptiles have nucleated red blood cells. Although amphibians do also, mammals, the only other endothermic vertebrates and the only other group with which birds might therefore be considered closely related, do not. Birds and reptiles also share a similar kidney structure and function, both excreting uric acid rather than the urea of mammals. Only reptiles and birds share a nasal salt gland for the excretion of concentrated excess body salts. The hoatzin, a peculiar bird from South America, has extraordinarily well-developed claws on its forearms that it uses to help navigate through the trees.

Studies of the development of reptiles and birds lead one to the unequivocal conclusion that the two groups share a common ancestry. One can actually see the fusion of various bones during embryological development in birds that remain separate in reptiles. For example, digits two and three are separate in reptiles and fused in adult birds, but one can find these bones separate in embryonic birds. After studying the development of the chicken embryo as far back as 1907, Frank Lillie concluded that the skull bones in reptiles and birds were essentially identical. They only look different in adults because the bones fuse tightly in adult birds, obliterating the embryonic suture patterns. Although many birds lack a penis as adults, there is sometimes embryonic penile tissue that is later resorbed. The penis in adult birds that possess one (e.g., geese, ducks, swans, ostrich, cassowary, rhea, kiwi) is basically identical in embryonic structure to that of adult reptiles. This organ is not a closed tube as in mammals, but rather has, in both reptiles and birds, a longitudinal furrow that guides the sperm during copulation. So it should come as no surprise that fossil evidence linking birds with reptiles would eventually become known. What did come as a surprise was the quality of the evidence. Before we get into this evidence, we need to look at the grand scheme of tetrapod and bird genealogy (phylogeny) and become familiar with some of the names of the higher taxonomic categories, because quite a few of them are mentioned in this chapter.

Genealogy of Dinosaurs and Birds

All four-footed vertebrated animals are classified in the Tetrapoda. This includes the modern amphibians and a host of other early four-footed animals, many of which were probably not on the line to living amphibians (see Chapter 10). Sometime late in the Devonian Period, terrestrial tetrapods developed the amniote egg, a structure with its own little ocean and protective membranes that could be laid on land instead of in the water. In one form or another, reptiles, birds, and mammals all have this egg and are therefore collectively classified as the Amniota. The Reptilia is a group of amniotes including all the known extinct and extant reptiles plus the birds. The next branch in the tetrapod family tree that concerns us is the Archosauria. This group includes the crocodiles and their extinct relatives (the Pseudosuchia) and the Ornithosuchia (the pterosaurs, dinosaurs, and birds). Pterosaurs (or pterodactyls) are the flying dinosaur-like archosaurs. They are considered as a sister group, or a branch related, to the Dinosauria (dinosaurs and birds). Once we get into the dinosaurs proper, paleontologists separate out the herbivorous forms, such as the giant *Brachiosaurus,* from the carnivores such as *Allosaurus, Tyrannosaurus, Velociraptor,* and *Coelophysis.* These are the Theropoda, or the theropods, and the group includes all the well-known carnivorous dinosaurs and the birds. As can be seen in **Figure 7.2**, a number of names further compartmentalize the theropods based on a variety of features, including size of the hands, number of digits, presence of a furcula ("wishbone"), position of pubis bone, number of caudal (tail) vertebrae, and finally a number of features defining birds (Aves), such as a specialized tail vertebra known as the pygostyle, a keeled sternum, and a reversed big toe for perching.

Amazing *Archaeopteryx*

Paleontologists refer to a gold mine of paleontological treasures, a locality with a fabulous hoard of fossils, as a "lagerstatten." The Burgess Shale, with its incredible representation of early Cambrian life, is such a place, and so is the Solnhofen limestone, though not necessarily for its quantity of fossils. The species name for *Archaeopteryx lithographica* derives from the nineteenth century use of the Solnhofen limestones in the lithographic printing industry, though it has also been used since Roman times to pave highways and build homes. Miners have been quarrying out almost perfect sheets of hard limestone for more than 100 years. The stone was apparently laid down in a very shallow possibly brackish water bay about 150 million years ago (Ma), probably under extremely anaerobic conditions, and whatever animals died there were quickly entombed. This is suggested by the incredible preservation of many fossils. Although fossil birds and reptiles are not common, when they are found they are often represented as entire articulated skeletons, and the imprint of their external anatomy often shows up in the hardened substrate around them. As related by Gerhard Heilmann, the first specimen of *Archaeopteryx* made its way in 1861 to the British Museum of Natural History in London, where it was described by the paleontologist Richard Owen, who also described the first dinosaur and the first land fossils from North America. Since that time six additional specimens have

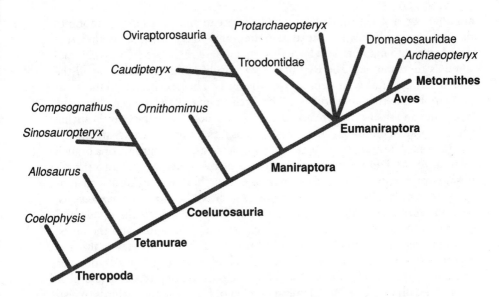

Theropoda
Thin bone walls; 40 caudals; glenoid faces posterolaterally; loss of manual digit V, reduction of IV; tibia > femur; long metatarsals; forelimbs < 40% of hindlimb

Tetanurae
Ossified sterna; boomerang-shaped furcula; hand reduced to digits I-III.

Coelurosauria
Hand is 2/3 - 3/4 humerus + forearm length; tuft-like or fringe-like integumentary structures (protofeathers?)

Maniraptora
Semilunate carpal; vaned, barbed feathers; coracoids fit in sternal grooves; 23-26 caudals

Eumaniraptora
Longer arms and hands; lateral glenoid in some taxa; forelimbs > 2/3 hindlimbs; pneumaticized pleurocoels; posterior migration of pubis

Aves
Arms = legs; lateral glenoid; longer fingers and forearm; < 23 caudals; flight feathers; airfoil; flight stroke

Metornithes
Arms > legs; fully reversed hallux; alula; pygostyle; hypocleideum; dorsolateral glenoid; sternal keel; etc.

Figure 7.2 A phylogenetic hypothesis for the relationships of dinosaurs and birds. Characteristics of the major groups are provided. (Adapted from K. Padian, 2001. Cross-testing adaptive hypotheses: phylogenetic analysis and the origin of bird flight. *American Zoologist* 41: 598–607.)

been discovered, some initially masquerading as other kinds of animals, particularly the second group of reptiles that evolved true flight, the pterosaurs. As we shall see, many fossil birds have been discovered that continue to link reptiles and birds, but none so spectacular as *Archaeopteryx*. We need to take a close look at this amazing animal.

A wondrous feature of some of the *Archaeopteryx* material is the minute preservation, as imprints, of the feathers. They have individually the exact structure as in modern birds. But their arrangement on the posterior region of the body is decidedly unlike birds, for they extend symmetrically from a long bony tail. In all modern birds the tail vertebrae are reduced to a single unique structure called the pygostyle, from which the tail feathers radiate. *Archaeopteryx* instead retains the multiple tail vertebrae of reptiles but with feathers instead of scales. This animal as an adult also had three free digits on the forearm rather than the partial fusion and elimination of some digits seen in most modern birds. The clavicles of *Archaeopteryx* were fused into a single structure called the furcula, as in modern birds. Perhaps the most astonishing feature of *Archaeopteryx* is the presence, in the upper and lower jaws, of complete and obviously functional rows of conical reptilian teeth (just recently, developmental biologists have been able to induce the development of primordial, though nonfunctional, teeth in chick embryos). The animal is so obviously a combination of reptilian and avian features that some scientists became suspicious that it was a fraud, like Piltdown Man. However, recent studies by Larry Martin and other reputable scientists have confirmed its authenticity. At 150 Ma, *Archaeopteryx* is the oldest known bird capable of dynamic flapping flight.

Reptiles, Specifically Dinosaurs, as Avian Ancestors

We have seen that many features are suggestive of an evolutionary tie between reptiles and birds, and *Archaeopteryx* provides us with first-order evidence, or a conclusive test, of that hypothesis. We have no way of influencing the outcome of what the fossils tell us; they are truth entombed in stone. The fossils could have told us another story; they could have falsified our reptile-bird hypothesis. Instead, they supported it. This is how science works, and there is no other tenable explanation for the similarities of birds and reptiles, together with the fossil evidence *Archaeopteryx* provides, than that birds evolved from reptilian ancestors. But which ancestors? Which brings us back to dinosaurs and other reptiles.

Three reptilian groups, the crocodylomorphs, extinct basal archosaurs (sometimes referred to as primitive "thecodonts") and the extinct coelurosaurian theropods have been postulated at various times as bird ancestors. Scientists even briefly toyed with mammals as possible bird ancestors, but this idea has been abandoned in recent years. The crocodylomorph hypothesis was promoted mostly by Alick Walker and Larry Martin and centered around a series of purported similarities between early crocodile ancestors and birds. As Walker showed, early crocodile ancestors of Triassic–Jurassic boundary age were rather lightly built agile predators, much like early dinosaurs. However, Walker later retracted his previous suggestions, favoring the theropod hypothesis, and the supposed shared derived characters ("synapomorphies") of crocodylomorphs and birds have since been shown to be either convergent features or more widely distributed among reptiles and therefore of no value for the purpose intended. Crocodilians are now considered an early branch of the archosaurian reptiles that are not on the line toward birds.

The hypothesis of a thecodont (basal archosaur) ancestry is a bit more complicated because both questions of direct ancestry and philosophy of classification play a part. The thecodonts, as currently understood, include a wastebasket of similar-looking primitive archosaurs. Many paleontologists do not recognize them as a distinct taxonomic group. There is little debate among paleontologists that all later archosaurs, birds, and crocodiles are descended from one or more of these species; the question is whether birds are descended *directly* from one of them or from a later theropod stage. Max Hecht and Samuel Tarsitano suggested that all known early birds, such as *Archaeopteryx*, occur too early in time and are too specialized to be descended from any known theropod dinosaur. They concluded that we are therefore forced to look among the earlier archosaurs for bird ancestry. One of the continuing problems with this idea is that there are no adequate fossil candidates. All known early thecodont species are likely to have been quadrupedal, rather than bipedal, and none has fused clavicles or a sternum.

Jacques Gauthier, Kevin Padian, and Luis Chiappe conducted a number of careful phylogenetic analyses with many characters from fossil reptiles and birds, showing that birds fit snugly within the coelurosaurian theropod dinosaurs (see Fig. 7.2, **Fig. 7.3**). Although it expresses some uniquely avian features, most of the skeletal characters of *Archaeopteryx* can be duplicated among the theropods. Some of these, which are not found in earlier generalized archosaurs, include the following and are taken from a recent summary by Ostrom:

1. Presence of fused clavicles;

2. Fibula complete, not fused with tibia;

3. Four-toed foot in which the outer toe (V) is lost and the inner toe (I; the hallux) is short and opposable;

4. Partially fused metatarsals with digit III as the longest;

5. Straight tibia with cnemial crest;

6. Nearly straight femur with well-defined greater and lesser trochanters;

7. Three-fingered hand with clawed digits, in which the middle digit is the longest;

8. Long and slender S-shaped humerus;

9. Strap-like scapula;

10. Subrectangular to semicircular coracoid;

11. Mostly amphicelous vertebrae;

12. Ilium and pubis oriented as in theropods;

13. Presence of gastralia (ventral abdominal ribs).

The Coelurosauria, as Gauthier defines the taxon, includes birds and a number of small tetanurine theropods that are more closely related to modern birds than they are to other theropods. They have long arms and hands and a variety of other skeletal features that are distinctly avian. Among this group is a subset, the *Maniraptora*, with even more specifically avian features, including all the accouterments for flight. For example, the wrist is composed of a special half-moon shaped bone that allows the hand to move in a broad arc, allowing typical flight motion. The *Maniraptora* include the modern birds as well as the extinct families Caenagnathidae, Dromaeosauridae,

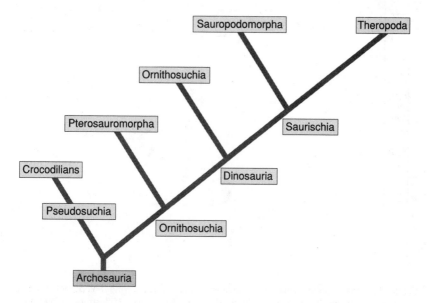

Figure 7.3 Phylogenetic hypothesis for Archosauria, including the Theropoda from which birds may have evolved.

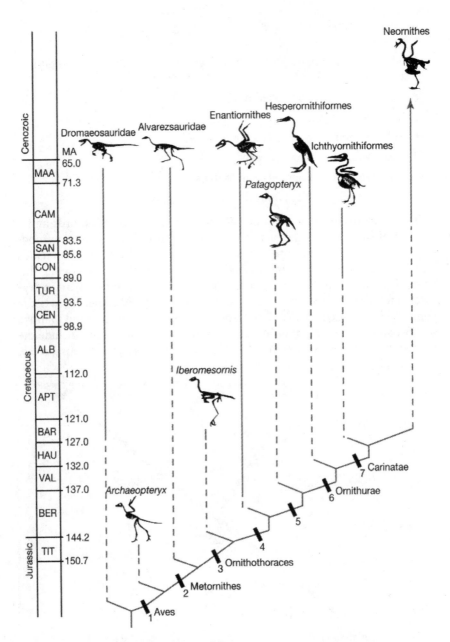

Figure 7.4 Phylogenetic scenario for extinct and extant birds, showing their potential close relationship to some gracile, carnivorous, bipedal dinosaurs. All of the birds illustrated except the Neornithes possessed sharp reptilian-like teeth, although one Cretaceous bird, *Confusciusornis*, had a beak with no teeth. (From L.M. Chiappe, 1995. The first 85 million years of avian evolution. *Nature* 378: 349–355. Copyright Nature Publishing Group.)

and Troodontidae (see Fig. 7.2, **Fig. 7.4**). This cladistic work is impressive and very convincing to many enthusiasts but not to Alan Feduccia and a few others, who see most of these similarities as convergent features. Feduccia is convinced that *Archaeopteryx* was an arboreal animal well on its way to becoming a modern bird and that we must therefore look to earlier probably unknown semiarboreal or arboreal thecodonts for ultimate avian ancestry.

Origin of Flight

The earliest serious speculations about the origin of flight date back to the late 1870s and early 1880s and were even then polarized into the two positions that are still hotly debated today. Othniel Marsh, a well-to-do paleontologist, famous for financially subsidizing early dinosaur excavations in the western United States, first proposed that birds evolved from an arboreal reptilian ancestor (the "trees down" hypothesis). In this model, the preavian reptile is seen as a tree-climbing limb-hopping animal, perhaps a primitive glider, with fringed scales on their way to becoming feathers. A volant (gliding) mode is presumed to have preceded true flapping flight. Extant gliding animals, such as flying squirrels, flying lemurs, *Draco volans* (a "parachutist" lizard), and others are often cited as examples.

Samuel Williston is credited with the first presentation of the cursorial ("ground up") theory, stating that birds were descended from cursorial (walking and running) dinosaurs. He also suggested that feathers evolved as an aid to running. This theory was elaborated more fully during the early 1900s by a colorful scientist-brigand from Transylvania, Baron Franz Nopcsa. He served as a spy in World War I, volunteered as monarch of Albania when the throne was vacant, and finally committed suicide after murdering his homosexual lover. From the remains of *Archaeopteryx* Nopcsa concluded much the same thing as Williston, that hand flapping by a terrestrial "pro-avis" stage intermediate between dinosaurs and birds aided running. He could not conceive of the obligatory bipedality of *Archaeopteryx* arising from anything other than a bipedal dinosaur.

Until recently there has been almost unanimous agreement that *Archaeopteryx* was an obligatory biped but probably an active but feeble flier. This conclusion seemingly follows from the structural similarities of *Archaeopteryx* to all of the bipedal theropods, including the larger carnivores such as *Tyrannosaurus* and *Allosaurus,* and to the absence of some anatomical features found to be well developed in modern flying birds, such as a distinct keel on the sternum where the major flight muscles originate (**Fig. 7.5**). The furcula and hand structure of *Archaeopteryx* have been shown to be common to some coelosaurian theropods and therefore are obviously not always associated with dynamic flight, unless one wishes to seriously suggest that *Velociraptor* flew. These observations led John Ostrom to develop a

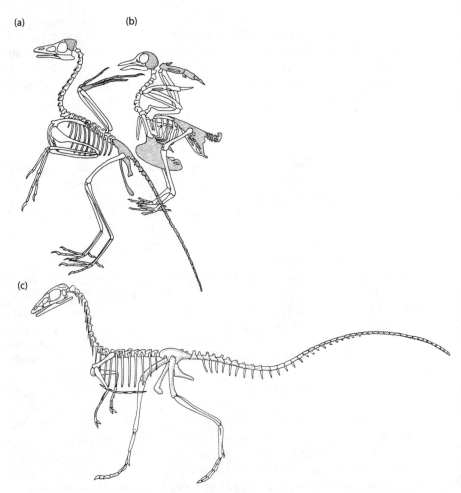

Figure 7.5 Comparison of the skeletons of *Archaeopteryx* (a), a modern bird (b), and the diminutive dinosaur *Compsognathus* (c). (From D.J. Futuyma, 1998. *Evolutionary Biology,* 3rd ed. Sunderland, MA: Sinauer Associates. p. 152.)

unique idea that he called the "cursorial predator" theory. This hypothesis states that feathers on the arms arose as an aid to capturing prey, either by batting it down from the air or surrounding it, as in a large snare. Ostrom sees the wings at this stage being used as a "flailing" device and the lift obtained by the wings as an added advantage in prey-catching behavior. A number of features of *Archaeopteryx* do in fact point to it as a likely predator, such as the sharp conical teeth and the sharp claws on a theropod-like foot. So, to Ostrom, *Archaeopteryx* was likely a terrestrial predator, taking to the trees only occasionally.

In a volume of published studies resulting from an international conference on *Archaeopteryx* held in Eichstatt, Germany in 1984, Richard Thulborn and Tim Hamley suggested an alternative to these terrestrial models, namely that flight could have originated from an aquatic form and the rudimentary wings could have carried the protobird from wave crest to wave crest. These scientists see *Archaeopteryx* filling the ecological role of today's larger predatory wading birds, such as egrets and herons. They further suggested that the feathered wings of *Archaeopteryx* might have been used as a canopy, shielding the sun to better reveal underwater prey, such as fish. This technique is used today among wading birds, particularly the African Black Heron. Sometimes the prey actually seek the shelter of the created shade, making them easy targets. Prey may also have been kicked up by the feet or by flicking motions of the tail. The plumage of *Archaeopteryx* is seen here both as a thermoregulatory and waterproofing device. In this scenario, *Archaeopteryx* is considered as at least a metaphorical structural ancestor to a swimming/diving bird that then developed powered flight by flapping from wave to wave. Updrafts from wavefronts might have assisted the bird into the air. The presence of water could have also served the useful function of providing protection during early experimental stages, cushioning a fall when the flight did not fare so well.

It is fascinating how the theories go around and around, but it is always new anatomical evidence and better fossils that turn the tide. In a recent publication Alan Feduccia showed that the claws of the front hand of *Archaeopteryx* were constructed like those of trunk-climbing birds and those of the hindfoot like perching birds. The evidence he presented was overwhelming; *Archaeopteryx* was arboreal. He also demonstrated that the asymmetrical contour feathers of *Archaeopteryx* are unique to modern birds and indicate that *Archaeopteryx* flew. Although they may have doubled as an insulating device to promote endothermy, the anatomical evidence indicates to Feduccia that the feathers of *Archaeopteryx* functioned for flight, not as a prey catching device. Even if Ostrom's cursorial predator hypothesis is correct, it did not apply to *Archaeopteryx* but to some earlier ancestral species.

Where does this leave us with regard to the origin of flight and the relationships of birds? The aquatic theory, interesting as it is, does not seem to have many adherents, primarily because of the many structural features shared by *Archaeopteryx* and terrestrial theropod dinosaurs plus the almost certain perching status of *Archaeopteryx*. As Kevin Padian noted in a 2001 article published in *American Zoologist,* we are not likely to ever know the answer to the "top down" versus "bottom up" controversy on the origin of flight, at least not framed in that manner. Flight can theoretically evolve either way. And the fossils still are not much help. The small coelurosaurian dinosaurs related to *Archaeopteryx* all occur in the fossil record after *Archaeopteryx* and so cannot be directly ancestral, although as Kevin Padian has shown, small delicate theropods existed in North America about the same time as *Archaeopteryx* in the Old World. However, no early generalized archosaurs are known that could be a common ancestor to both dinosaurs and birds, so the dissident camp, those that are against a dinosaur-bird (ground-up) ancestry, cannot point to obvious bird ancestors either. And it may be that feathers evolved initially neither for flight nor for predation. Maybe they evolved primarily in males to attract mates and were enhanced because females mated preferentially with the more gaudy individuals. This would explain their presence in both nonflying dinosaurs and birds. This evolutionary mechanism, known well to Darwin and subsequent generations of biologists as sexual selection, is considered the current explanation for many bizarre structures in animals, from the tails of swordtail fishes to the antlers of deer and elk. The feathers would, of course, have also provided a thermoregulatory advantage (perhaps first in the newborn; consider down feathers), and the eventual use of wings to create lift would then have opened new avenues for feeding and migration. Does this favor the ground-up model and a dinosaur-bird ancestry? Possibly, and given the many careful cladistic studies that seem to link birds and dinosaurs, the evidence seems to be slightly shifted in that direction for now (**Fig. 7.6**).

Frauds and Feathered Dinosaurs

There have been only two great frauds in the history of paleontology, Piltdown Man (see Chapter 11) and "Piltdown Bird," as the 1999 *Archaeoraptor liaoningensis* is sometimes called (a third, minor fraud, centering around some Himalayan invertebrate fossils and an Indian paleontologist, will not be considered here). The first was an elaborate hoax, but the second was likely either an accident or an attempt by a poor Chinese peasant to make a little extra money. Nevertheless, the story behind *Archaeoraptor* is worth repeating, if for no other reason than to demonstrate how science works at both its worst and its best and eventually corrects its mistakes.

The Liaoning region of northeastern China is fast becoming famous for its production of important dinosaur and early bird fossils and just recently also revealed a skeleton of what may be the earliest placental mammal. Feathered dinosaurs, such as *Sinornithosaurus* and *Beipiaosaurus*, had previously been described from deposits in Liaoning Province dated to around 125 Ma. So we already knew that feathers had not evolved only in birds and only for flight. But the scientific world became excited by what appeared to be another "missing link" in the reptile-bird transition, reported in a 1999 article in *National Geographic Magazine* by Christopher Sloan, then a Senior Assistant Editor at National Geographic. In his report, "Feathers for *T. rex*?", Sloan described a fossil that appeared to combine a bird's body and limb

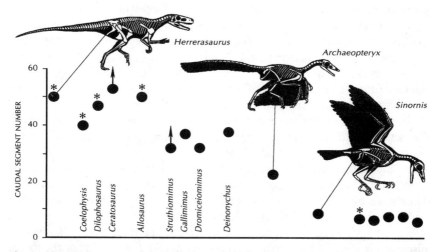

Figure 7.6 Graph of number of caudal segments (usually vertebrae) on *y* axis against time on *x* axis. Arrows indicate that there were likely more segments than were preserved; asterisks indicate estimates of full caudal component. Five dark circles at bottom right are advanced birds. (From S.M. Gatesy, and K.P. Dial, 1996. Locomotor modules and the evolution of avian flight. *Evolution* 50: 331–340.)

anatomy (including imprints of feathers) with a dinosaur's (specifically a dromaeosaur's) straight tail. Because the name of the fossil, *Archaeoraptor liaoningensis,* first appeared in that article, Sloan's name will always be associated with the beast, despite the fact that he did not discover the animal, prepare it for study, or examine it in any professional way. In one of those strange and ironic twists that often occur in real life, the professional scientists who jockeyed for this honor and the potential attendant fame were saved from this embarrassment by the scientific review process and the haste of National Geographic staff to publish a hot story. Much of what follows is taken from the careful detective work of Lewis Simons, who was hired by *National Geographic Magazine* Editor Bill Allen to investigate the *Archaeoraptor* scandal with the mandate to "let the chips fall where they may."

In July 1997 a farmer, digging in a shale pit, discovered a bunch of fossils in a number of shale slabs. In these layered sediments, sometimes part of a specimen can be on one side and part on another side of a slab; to reveal the fossils one has to actually split a slab lengthwise, as in opening an Oreo cookie, to use Simons's apt analogy. One side is called the "part" and the other the "counterpart." On that day the farmer had found pieces containing at least two kinds of animals, one a dromaeosaurid dinosaur and the other a primitive bird. The farmer trudged back to his home in a small village and glued the specimens together in ways to make them appealing to a dealer who would pay him a small sum for them. Simons was unable to determine if the specimens forming *Archaeoraptor* were combined by accident or purposefully. He was not able to speak with the farmer, as trading in the fossil black market in China is illegal and the farmer could have faced stiff penalties, perhaps even execution, if he was identified. Simons did speak with the dealer, who admitted he sometimes sold "composite" specimens. The dealer smuggled the *Archaeoraptor* specimen out of China, where it ended up in a gem and mineral show in Tucson, Arizona.

The *Archaeoraptor* specimen was initially seen at the show by Steven Czerkas, director of a small nonprofit museum in Blanding, Utah. He purchased the specimen for $80,000 with funds provided by M. Dale Slade, a Blanding businessman. Czerkas and his wife Sylvia are very knowledgeable dinosaur enthusiasts who have written papers and books on these animals. They were, quite understandably, extremely excited about the new specimen and were planning on making it the focal attraction at their little museum. Within about a week after returning to Utah with the specimen, the Czerkases discussed it with Philip Currie, a renowned dinosaur specialist at the Royal Tyrrell Museum of Paleontology in Alberta, Canada. They invited Currie to join them as an author on a paper describing the specimen. Currie then alerted Christopher Sloan at the National Geographic of the find, and Sloan thought there was a potential story for the magazine. Because the *Archaeoraptor* specimen would undoubtedly be considered by the Chinese as a smuggled national treasure, Currie convinced the Czerkases that it should be returned to them following study. The Czerkases agreed. Also at Currie's suggestion, Xu Xing, a scientist at the Institute of Vertebrate Paleontology and Paleoanthropology, was dispatched to the United States to contribute to the study. He spent 2 days in this examination in Blanding and then flew to a news meeting in Washington where the find was announced to the media.

Sloan, an art director, obtained Bill Allen's permission to cover the *Archaeoraptor* story and eventually write the piece, his first for National Geographic. Meanwhile, Philip Currie had finally seen the specimen on March 6. He says he did not like the look of the body-tail connection, but neither he nor the Czerkases can agree on what was said about it to each other, and by this time Bill Allen had thrown a net of secrecy over the project, which limited communication between the active parties. In May, Sloan looked at the specimen in Utah, but not being a scientist he did not see anything unusual and nobody suggested to him there might be a problem. On August 2, Currie met the Czerkases in Austin, Texas at Tim Rowe's x-ray facility. Rowe is a well-known paleontologist at Texas who specializes in high-tech approaches to the study of fossils. He has produced, for example, some outstanding computed tomographies of important fossil skulls. Rowe scanned the *Archaeoraptor* specimen and concluded there were no fewer than 88 breaks on it. Some of the fractures appeared to be between pieces that did not match well. He told Simons that he had explained to Currie and the Czerkases there was a chance the specimen was a fraud. However, regardless of who told what to whom, after they left the scanning facility they submitted a description of *Archaeoraptor* to the prestigious journal *Nature,* which, although it has a standard peer review process, can usually publish important articles pretty quickly. To their credit, it was everybody's intention, including the National Geographic staff, to have the *Archaeoraptor* specimen described and named in the scientific literature before the report of it came out in the National Geographic magazine (the magazine was to go to press on September 19). But it never happened.

In the first week of September, Currie sent Kevin Aulenback, a fossil preparator at the Tyrrell Museum, to Blanding to prepare the specimen for better examination and illustration. Aulenback alerted both the Czerkases and Currie that the specimen was likely a composite. However, he did not relay this concern to the National Geographic staff, although at one point he reported on his preparation to Sloan. As noted above, earlier in August Currie, Rowe, Hu, and Steven Czerkas had submitted a description of *Archaeoraptor* to *Nature* in London. A copy went to Sloan in Washington. Even in the paper, the authors noted the mosaic nature of the specimen, but it still did not sound an alarm at National Geographic headquarters. Henry Gee, the senior editor at *Nature,* indicated to the National Geographic staff that the paper could not be reviewed and published in time for the September 19 deadline. The scientists dashed off a modified version of the paper to the American counterpart of *Nature, Science* magazine. It was reviewed quickly (twice) and just as quickly rejected from publication. This also should have sent up some red flags at National Geographic, but it did not. They decided to go ahead with the September 19 printing.

On December 20 Xu Xing contacted his coauthors with the really bad news; in China he had been shown the unbroken counterslab of the *Archaeoraptor* tail, and it was connected to a dromaeosaur body. In March 2000, National Geographic published a letter to them from Xu indicating the original *Archaeoraptor* specimen was a composite, and in October 2000 Simons wrote his investigative account. So what do we conclude from all of this? A quick search for *Archaeoraptor* on the Internet demonstrates that Christian fundamentalists tried to argue from the fraud that missing links do not exist. But this just demonstrates that fundamentalists do not understand what missing links are in the first place. There is not going to be one fossil specimen that is exactly half way between a perfect dinosaur and a typical bird, but rather a series of intermediate steps within a number of different species, stretching over at least thousands if not millions of years. And because mosaic evolution seems to dominate in the early history of successful animal radiations, we can expect to see all kinds of dinosaurs trying and not quite succeeding at making the transition to birds (and yet being successful during their own time). If the fossil record of feathered dinosaurs is any indication, we are just seeing a very small proportion of the feathered reptilian diversity of the middle and later Mesozoic. I believe if we dropped in on the late Jurassic of northeastern China we might find it a lot gaudier and noisy place then we could ever imagine.

It is a testimony to the scientific process that only two frauds have occurred in more than 200 years of paleontological research and both have been uncovered by scientists and people who are interested in the truth, as Simons said, wherever the chips may fall.

Beyond *Archaeopteryx*

Because this is not a chapter on the evolution of dinosaurs, I am not going to review here all the many fascinating middle and late Mesozoic carnivorous theropods, maybe most of which had feathers or filamentous feather-like structures at some point in their lives. Some of these include *Longisquama* from Kyrgyzstan, *Unenlagia* from South America, *Coelophysis* from New Mexico, and *Caudipteryx, Sinornithosaurus,* and *Protarchaeopteryx* from the Liaoning deposits of China, not to mention the many carnivorous species related to *Tyrannosaurus, Velociraptor,* and *Deinonychus.* If birds evolved from dinosaurs, then all of the gracile, light-bodied, ornithomimid and troodontid dinosaurs are apparently on the line leading to birds.

The earliest undisputed bird is still the late Jurassic *Archaeopteryx,* at about 150 Ma. Almost all other Mesozoic birds are Cretaceous in age (some from China are possibly latest Jurassic, but this is not yet clear), showing that by that time a major avian radiation was in full swing. Bird-like animals beyond *Archaeopteryx* are sometimes classified in a supergroup called the Pygostylia, indicating that they possessed a specialized last caudal vertebra that supports, in living birds, both the tail feathers and a preen gland, a sebaceous (oil producing) gland used by birds to protect their feathers. We do not know when the first preen gland appeared in the Pygostylia, but it is not unreasonable to assume it may have evolved about the same time as the pygostyle (**Fig. 7.7**). The earliest record of a beaked bird is *Confuciusornis* from the Yexian deposits in Liaoning Province, China. The tail in *Confuciusornis* is reduced to four to five free caudal vertebrae plus the pygostyle. As in *Archaeopteryx,* the forearm includes three free clawed digits, the tarsometatarsus is not fused distally, gastralia extend posteriorly from the sternum, and metatarsal V is present. The skull of *Confuciusornis* is a curious mixture of primitive and advanced traits. Whereas the anterior part, with loss of teeth and a horny beak, is advanced, the posterior part

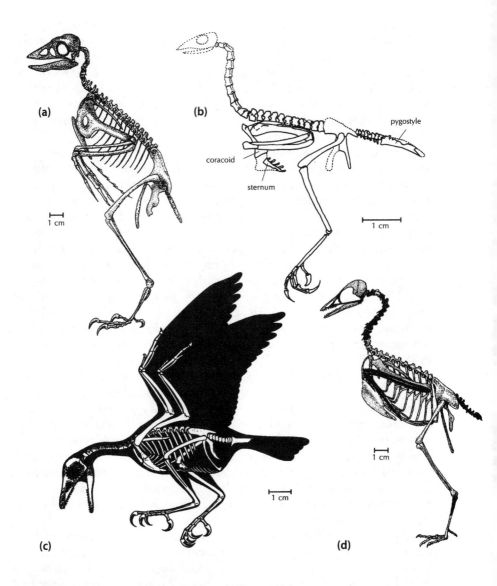

Figure 7.7 Cretaceous birds. (a) *Confuciusornis* from China, (b) *Iberomesornis* from Spain, (c) *Sinornis* from China, and (d) *Chaoyangia* from China. ([a] Reprinted from *Nature* [Lian-hai Hou et al., vol. 377]; copyright 1995, Macmillan Magazines Limited; [b] J.L. Sanz, and J.F. Bonaparte, 1992. A new order of birds (Class Aves) from the Lower Cretaceous of Spain. *Los Angeles Co Mus Nat Hist Sci Ser* 36:39–50; [c] Reprinted with permission from P.C. Sereno, and R. Chenggang, 1992. Early evolution of avian flight and perching: New evidence from the Lower Cretaceous of China. *Science* 255: 845–848; [d] Reprinted with permission from L. Hou, L.D. Martin, Z. Zhou, and A. Feduccia, 1996. Early adaptive radiation of birds: Evidence from fossils from Northwestern China. *Science* 274: 1164–1167. Copyright 1992, 1996, American Association for the Advancement of Science.)

retains the primitive theropod diapsid (two holed) condition. Fossilized imprints of this 120 Ma bird show that some specimens had two long graceful feathers extending from the tail, as in some modern flycatchers. According to Larry Martin and colleagues, only about 5% to 10% of the *Confusciusornis* specimens have these feathers, and if this is not an artifact of deposition, then these are likely males. Also, many of these birds are found together in the fossil deposits, making this the earliest likely record of flocking behavior in birds.

As in *Archaeopteryx,* almost all the Mesozoic birds (Aves) are differentiated from their modern relatives by virtue of the manner by which fusion of the tarsal elements develops in young individuals, from proximal to distal, rather than in the reverse direction. These early birds are classified together in the Enantiornithes, whereas the Neornithes includes the modern birds. Most of the enantiornithines possessed teeth, as in reptiles, but were in other features distinctly avian, though many did not fly. Some of the best known enantiornithines include *Liaoningornis, Eoenantiornis, Liaoxornis, Cathayornis, Eocathayornis,* and *Baluochia* from the early Cretaceous of China; *Gobipteryx* from Mongolia; *Iberomesornis, Concornis, Noguerornis,* and *Eoalulavis* from early Cretaceous deposits in Spain; and *Rahona* (if it is not an accidental composite of a bird and a dinosaur) from the late Cretaceous of Madagascar. The interesting thing about *Rahona,* described by Catherine Foster and colleagues in 1998, is that it supposedly has a slashing sickle-shaped claw on its feet as in *Deinonychus. Sinornis cantensis,* described by Paul Sereno and P. Chenggang from the early Cretaceous of China, was a sparrow-sized enantiornithine that inhabited the shores of a freshwater lake. The hand and wrist elements, as in *Archaeopteryx,* were unfused (fused in modern birds), though the third digit was more reduced than in *Archaeopteryx.* The sternum was ossified but unkeeled. The tail vertebrae were reduced to the pygostyle of modern birds. A chain of false ribs (gastralia) was retained, as was the primitive theropod form of the pubis. As in *Archaeopteryx,* the bill included rows of sharp reptilian teeth. But a number of features show that, despite the lack of a keel on the sternum, *Sinornis* was capable of fully powered flight.

At the same time the enantiornithines were dominant, another group more closely related to modern birds, the ornithurines, also put in an appearance. The Ornithurinae includes a variety of specialized extinct toothed birds such as *Hesperornis* from the Cretaceous of North America, a large aquatic bird with diminutive wings and outsized probably webbed feet, and *Ichthyornis,* also from the North American Cretaceous, that flew and likely ate fish, filling the role of today's cormorants. These birds were recovered from freshwater deposits in the western United States, and recently another ornithurine, *Halimornis,* was described from late Cretaceous marine sediments in Alabama. Other ornithurines are represented in the early Cretaceous Yixian Formation

Figure 7.8 *Presbyornis,* a genus from the Paleocene and Eocene of North America, showing transitional skeletal anatomy between a stilt-legged shorebird and a duck. (From S.L. Olson, and A. Feduccia, 1980. *Presbyornis* and the origin of the anseriformes (Aves: Charadriomorphae). *Smithsonian Contributions to Zoology* 323: 24.)

deposits in the Liaoning Province of northeastern China by *Liaoningornis* and *Chaoyangia*. Although ornithurines are as ancient as the enantiornithines, at least from the fossil record it appears that the toothy enantiornithines dominated the late Mesozoic, and it was not until after the late Cretaceous mass extinction event that modern birds, like modern mammals, diversified into their extant families.

There is further information in the avian fossil record of missing links, but at a different scale—the connection of ancient true birds with their modern counterparts. Alan Feduccia and other ornithologists have long supposed that most modern birds, other than the ratites and tinamous, were descended from shore birds. For instance, the body form and habits of modern ducks and flamingos are quite distinct from modern shore birds such as stilts and avocets, but this was not always the case, as the fossil genera *Presbyornis* and *Juncitarsus* demonstrate (**Fig. 7.8**). *Presbyornis,* known from the Paleocene and Eocene of western North America, combines the typical head and beak structure of a duck with the skeleton of a shorebird, whereas *Juncitarsus,* an Eocene taxon, demonstrates the head and unique forearm structure of flamingos coupled with the skeleton of a generalized shorebird. So all these ancient bird groups, combining as they do both reptilian and avian features, represent missing links on a grand scale, between what are currently considered two different classes of vertebrates.

It should be obvious to the reader by now that not one species represents the missing link between reptiles and birds but rather many species of protobirds, all experimenting with a preavian mode of life and displaying various combinations of reptilian and avian features. These experiments probably took place over millions of years. It is true that there was only one tree of life and that some of the species discussed in this chapter were possibly on the line to modern birds and some were not, but it is not important that we know exactly which species were ancestral. The fossil evidence simply cannot be explained except by an evolutionary process. First we have theropods with sharp claws, teeth, and no feathers; then theropods with feathers; followed by theropods with feathers and a beak; and finally modern birds. What other reasonable hypothesis explains these data?

Suggested Readings

Chiappe, L. M. 1995. First 85 million years of avian evolution. *Nature* 378:349–355.

Chiappe, L. M., J. P. Lamb, Jr., and P. G. P. Ericson. 2002. New enantiornthine bird from the marine Upper Cretaceous of Alabama. *Journal of Vertebrate Paleontology* 22:170–174.

Feduccia, J. A. 1994. Tertiary bird history: notes and comments; pp. 178–189. In D. R. Prothero and R. M. Schoch (eds.), *Major Features of Vertebrate Evolution*. University of Tennessee Press, Knoxville.

Feduccia, J. A. 1996. *The Age of Birds* (second edition). Harvard University Press, Cambridge.

Forster, C. A., S. D. Sampson, L. M. Chiappe, and D. W. Krause. 1998. The theropod ancestry of birds: new evidence from the Late Cretaceous of Madagascar. *Science* 279:1915–1919.

Gauthier, J. A. and K. Padian. 1989. The origin of birds and the evolution of flight; pp. 121–133. In S. J. Culver (ed.), *The Age of Dinosaurs*. University of Tennessee Press, Knoxville.

Hecht, M. K., J. H. Ostrom, G. Viohl, and P. Wellnhofer (eds.) 1985. *The Beginnings of Birds: Proceedings of the First International* Archaeopteryx *Conference*. Freunde des Jura Museum, Eichstatt, West Germany.

Lim, J.-D., Z. Zhou, L. D. Martin, K.-S. Baek, and S.-Y. Yang. The oldest known tracks of web-footed birds from the lower Cretaceous of South Korea. *Naturwissenschaften* 87:256–259.

Martin, L. D., Z. Zhou, L. Hou, and A. Feduccia. 1998. *Confusciusornis sanctus* compared to *Archaeopteryx lithographica*. *Naturwissenschaften* 85:286–289.

Olson, S. (ed.). 1999. Avian paleontology at the close of the 20th century. *Proceedings of the 4th International Meeting of the Society of Avian*

Paleontology and Evolution, Washington, DC, 4–7 June 1996. Smithsonian Institution Press, Washington, DC.

Ostrom, J. A. 1994. On the origin of birds and of avian flight; pp. 160–177. In D. R. Prothero and R. M. Schoch (eds.), *Major Features of Vertebrate Evolution.* University of Tennessee Press, Knoxville.

Padian, K. (ed.). 1986. The Origin of Birds and the Evolution of Flight. *Memoirs of the California Academy of Sciences,* No. 8.

Padian, K. 2001. Cross-testing adaptive hypotheses: phylogenetic analysis and the origin of bird flight. *American Zoologist* 41:598–607.

Sereno, P. and C. Rao. 1992. Early evolution of avian flight and perching: new evidence from the lower Cretaceous of China. *Science* 255:845–848.

Zhou, Z. 2002. A new and primitive enantiornithine bird from the early Cretaceous of China. *Journal of Vertebrate Paleontology* 22:49–57.

Zhou, Z. and L. Hou. 1998. *Confuciusornis* and the early evolution of birds. *Vertebrata Palaeontologica Asiatica* 36:136–146.

8

On Hearing
and Hinges

*". . . the next time you listen to Mozart or the Rolling Stones,
say a brief thank you to a long sequence of vertebrates that
made the successful transition from water to land. . ."*

The human hearing apparatus is a complex and amazing thing. Sound waves
are funneled by the ear, or pinna, and travel through a canal to cause a thin
membrane, the tympanum (tympanic membrane), to vibrate. These initially
airborne vibrations are then enhanced and transferred by three bones of the
middle ear to a structure of the inner ear known as the cochlea, a tightly
wound spiral structure. Vibrations set up in the liquid of the cochlea are
"read" by special cells with tactile ends, and the specific nerve impulses are
then speedily carried by the auditory nerve to the auditory region of the brain,
where they are processed and understood as sound. This basic structural pattern is found in all mammals, though there is considerable variation in auditory function. Some mammals, for example bats and whales, specialize in
high frequency sounds, although they are also capable of making lower frequency sounds audible to humans. Only mammals have three middle ear
bones: the malleus, incus, and stapes (**Fig. 8.1**). In addition, in mammals another bone, known as the tympanic, helps support the tympanic membrane.
Living amphibians, reptiles, and birds have just a single middle ear bone, the
stapes.

The efficient and sensitive hearing of mammals is a legacy inherited from
a distinct branch of ancient reptiles that first put in an appearance about 320
million years ago. They are classified together with mammals as the taxon
Synapsida, to distinguish them from other reptiles and birds, which are classified together in a "sister" group called the Sauropsida (sometimes also
known as the Reptilia; depends on the investigator). The synapsids include a
little-derived group of ancestral reptiles called the pelycosaurs and the more
advanced therapsids. Mammals evolved from the cynodont therapsids. Among

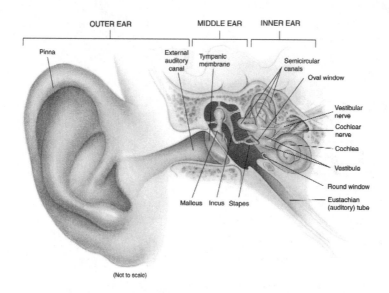

Figure 8.1 Ear structure in human beings. Note the three middle ear bones, the malleus, incus, and stapes. Reptiles have only the stapes, and our malleus and incus evolved from the reptilian articular and quadrate bones, respectively. (Marsha J. Dohrmann.)

other features, the Synapsida are allied by the presence of an opening in the skull behind the orbit known as the *lateral fenestra*. This opening becomes much enlarged in mammals as the muscles for chewing become better developed. The transition from mammal-like reptiles to mammals is one of the best documented evolutionary progressions in the fossil record and also establishes the homology and modification of bones of the lower jaw of ancestral mammal-like reptiles into the secondary bones of the mammalian middle ear.

Reptile–Mammal Transition

Energy Relations

Before we delve into a brief overview of synapsid evolution we need to consider some basic physiology of reptiles and mammals, because this will have some bearing on how we interpret the fossils, both in their classification and life-style. Living reptiles are *ectothermic,* meaning they derive the heat they need for biological activity from a source external to their bodies. More specifically, modern terrestrial reptiles must heat up to a particular level before the enzyme systems that drive their activity can function at maximum efficiency. They usually bask in the sun, which makes them *heliothermic* as well (helios = sun, thermic = heat). Living reptiles are also *poikilothermic,* meaning that their body temperature changes as a function of the external, or ambient, temperature. Because of this physiological mode, there are virtually no reptiles in the arctic and there are no reptiles active during the winter months in temperate and northern latitudes. Furthermore, living reptiles generally cannot sustain long bouts of aerobic metabolic activity; ergo, they cannot run long distances at high speeds.

Despite these limitations, reptiles can actually maintain fairly constant body temperatures throughout the day by behavioral means; they shuttle back and forth in and out of the sun and also may have the ability to change color—dark to take up heat and light to reflect it. The ability to maintain a constant body temperature is referred to as *homoiothermy*. Keeping a constant body temperature by external means is called *ectohomoiothermy*.

Mammals are more energetically efficient and therefore are *endothermic*. That is, their metabolic rate is much higher than that of the average reptile, and they are able to generate a high level of internal heat merely from internal combustion (metabolism). They are thus independent of external temperatures for activity. This gives mammals a decided edge in that they can exploit energy resources in colder environments and also during the winter. Additionally, mammals, like birds (the other living endotherms), have insulation to minimize heat loss. The combination of higher metabolic rate and insulation makes mammals *endohomoiothermic*. There is a price to pay for endothermy though; a mammal must consume at least six times as much energy at the same body size as a reptile.

It is often difficult to determine if a fossil species was ectothermic or endothermic based solely on its skeletal structure, and therefore the group Mammalia is not defined on this basis. Nevertheless, we can use certain features of the skull, such as a complete secondary palate, and the skeleton, such as joint–muscle relationships indicating highly efficient running, to imply that endothermy was likely. As we shall see, all mammals are endothermic, but all reptiles are not ectothermic.

What Is a Mammal?

If endohomoiothermy is not used to diagnose a mammal, then what is? Some scientists would, in fact, like to define mammals on the basis of endothermy, but it may be that this feature developed independently in a few lines of therapsids, some of which did not lead to mammals. Remember, to diagnose a group we must identify distinct synapomorphies for it, derived characters shared among members and with no other groups. The best feature for defining the monophyletic group Mammalia appears to be the way the lower jaw articulates with the skull (**Fig. 8.2**). In reptiles this is always the quadrate bone of the skull and the articular bone of the lower jaw. In mammals, the squamosal bone of the skull articulates with the dentary of the lower jaw. This is a par-

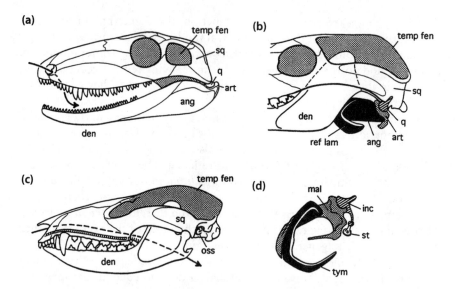

Figure 8.2 Skull, lower jaw, and ear structure in an Early Permian synapsid reptile *Varanosaurus* (a), a living primitive mammal, the Virginia opossum *Didelphis* (b), and the Early Triassic cynodont (mammal-like) reptile *Thrinaxodon*. Arrows indicate path of air. In (d) we see a magnified view of the opossum ear bones. The opossum has a secondary palate, the early synapsid does not. The shading in C and D indicates homologous elements in the middle ear region of *Thrinaxodon* and the opossum. Sq = squamosal, q = quadrate, art = articular, ang = angular, den = dentary, tym = tympanic, mal = malleus, inc = incus, st = stapes, temp fen = temporal (lateral) fenestra. (From J.A. Hopson, 1994. Synapsid evolution and the radiation of non-eutherian mammals. In D.R. Prothero and R.M. Schoch, eds. *Major Features of Vertebrate Evolution. Short Courses in Paleontology.* Lawrence, KS: The Paleontological Society. p. 192.)

ticularly good structural system to work with to define mammals because the fates of the quadrate and articular, as well as the reflecting lamina of the angular, are intimately tied to this new articulation and become the ear bones only in mammals. We look at this in detail in a section below, after we briefly examine the evolution of the synapsids. Additional characters sometimes used to define mammals include divided roots on teeth posterior to the canine and four, rather than three, incisors in each half of the lower jaw.

Early Synapsids

With scales instead of fur, early synapsids were almost surely ectothermic. One pelycosaur, *Dimetrodon*, had a large dorsal "sail" that extended off its back that was likely used as a thermoregulatory device, picking up heat from the sun in the morning. Pelycosaur teeth were rather simple and similar throughout the jaws, indicating that animal food was likely held in place and then bolted down but not processed as in mammals. Lower metabolic efficiency in early synapsids can also be seen in the structure of the palate, or roof of the mouth. In mammals the oral cavity is separate from the nasal cavity by a secondary shelf of bone known as the secondary palate. This shelf was absent in pelycosaurs, which would have created problems in breathing if chewing occurred. The limbs of early synapsids were also splayed laterally, and the body moved by undulatory motion, rather than being propelled quickly with the limbs under the body as in later therapsids and mammals. The quadrate bone of the skull formed a joint with the articular bone of the lower jaw, as it does in all other reptiles. More importantly for our consideration, the lower jaw was composed of many separate bones, of which the dentary was the largest and most anterior. The angular bone of the lower jaw of pelycosaurs was already slightly modified in the direction of mammals, possessing a curved structure known as the reflected lamina, which in later therapsids may have supported the tympanic membrane and which in mammals becomes the tympanic bone.

Therapsids

Later synapsids, the therapsids, demonstrate progressively more mammalian features. There is a reduction in number and size of bones in the skull and jaws, and the dental battery becomes more complex, indicating a more diverse diet than in earlier species. The posterior roof of the skull narrows, and the lateral fenestra enlarges while the zygomatic arches (cheekbones) flare out as the masseter muscles expand on the sides of the skull and lower jaws for more powerful mastication. The secondary palate develops. Pitting of the facial bones suggested to some investigators that later therapsids possessed whiskers, and thus fur, but such pitting is also found in modern reptiles and so the nature of the skin in therapsids remains unknown. The limbs become perpendicular below the torso, and all the bone–muscle articulations imply, in later carnivorous cynodont therapsids, highly efficient and mammalian-type locomotion and metabolism.

The earliest therapsids include a variety of carnivorous and herbivorous species arranged into the Biarmosuchia, Eotitanosuchia, Dinocephalia, and Anomodontia. Their remains are especially common in deposits of Russia and South Africa. They share many of the features of the primitive synapsids with more mammalian features and some specialized ones, indicating that many of them were not ancestral to mammals. Progressive features include enlargement of the lateral fenestra, more distinct upper canines, and a postcranial skeleton more indicative of efficient locomotion.

The more advanced therapsids include the Gorgonopsia, Therocephalia, and Cynodontia. Collectively they are known as the Theriodontia. The name Eutheriodontia is used for the Therocephalia and the Cynodontia taken together. Gorgonopsians were large predatory therapsids, lightly built and with

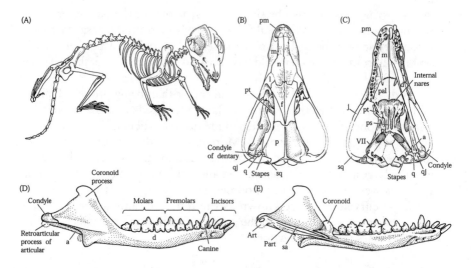

Figure 8.3 Skull, jaw, and skeletal morphology of an Early Jurassic mammal, *Morganucodon*. In one of those rare instances of outstanding preservation in the fossil record, we can see that *Morganucodon* is in transition from the reptilian to the mammalian condition, having retained not only some of the reptilian bones of the lower jaw but also the mammalian squamosal/dentary skull-jaw articulation; a true "missing link." (From D.J. Futuyma, 1998. *Evolutionary Biology*, 3rd ed. Sunderland, MA: Sinauer Associates. p. 151.)

serrations on the anterior teeth, the incisors, as well as large dagger-like canines. They were the *"Tyrannosaurus"* among the therapsids but were too highly specialized to be ancestral to mammals.

Cynodonts, including mammals as a subgroup, are considerably different from other therapsids from their first appearance in the Late Permian. *Procynosuchus*, widespread throughout east and southern Africa as well as Europe, is a good candidate for a common ancestor of all later cynodonts. Cynodonts possess extra cusps on the postcanine teeth, implying a greater processing of food. A partial secondary palate is found in early members of this group. The zygomatic arches are flared, and there is circumstantial evidence for the mammalian masseter muscle. The dentary is enlarged and extends more posteriorly. The vertebral column is distinctly divided into thoracic (chest) and lumbar regions. In *Thrinaxodon*, from the early Triassic, the secondary palate becomes complete. In this genus the cheek teeth are also more complex than those of *Procynosuchus*. Among the advanced cynodonts, James Hopson at the University of Chicago believes the mammals arose from near the trithelodontids (= ictidosaurs), such as *Pachygenelus*, with the Probainognathidae (e.g., *Probainognathus*) as close relatives.

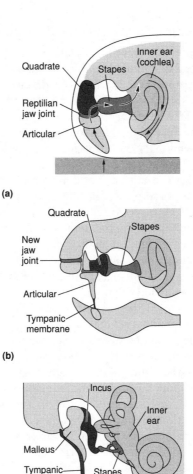

(a)

(b)

(c)

Figure 8.4 Evolution of the middle ear bones and the jaw joint during the transition from synapsid reptiles to early mammals. In early synapsid reptiles, (a) only one middle ear bone, the stapes, is present, and the skull articulates with the lower jaw by the quadrate (in skull) and articular (in jaw) bones. In the earliest mammals, such as the extinct *Morganucodon,* (b) the quadrate connects with the stapes and the articular, but a new joint evolved between the skull and the jaw. As in modern mammals, there is a connection between the squamosal bone of the skull and the dentary bone of the lower jaw. In living mammals (c) with more refined chewing and hearing, the articular bone has become the malleus and the quadrate the incus. (Adapted and modified from K.A. Kermack, and F. Mussett, 1983. The ear in mammal-like reptiles and early mammals. *Acta Palaeontologica Polonica* 28: 147–158.)

Evolution of the Hearing Apparatus and the Earliest Mammals

The oldest fossil specimen identified as a mammal is *Adelobasileus* from the Late Triassic of Texas at about 225 million years ago. It, and some other early specimens from Europe, are only tentatively defined as mammals because the jaw joint is not preserved. The earliest taxon for which adequate material is known is *Sinoconodon* from the Early Jurassic of China and *Morganucodon* of Europe. Showing once again that evolution progresses in small increments, we can see in these animals almost perfect intermediates between "reptiles" and "mammals." (I use quotation marks here because it should be apparent that "mammals" are specialized cynodont reptiles with fur and a new jaw joint.) Rather than having lost the therapsid skull–jaw articulation, in *Sinoconodon* and *Morganucodon* this reptilian joint lies right along side the new mammalian one (**Fig.8.3**). *Sinoconodon* retains multiple replacement sets of teeth, including the "molars" (a reptilian pattern), whereas in *Morganucodon* the replacement is limited to a single "milk" set, as in modern mammals, and the upper and lower teeth fit together (occlude) more precisely. Postcanine teeth have divided roots in both taxa.

The reflected lamina of the angular bone in *Morganucodon* is in a form that may have already supported the eardrum (tympanic membrane) (**Fig.8.4**). The small quadrate bone greatly resembles the incus of modern mammals. But it is evidence from the development of a modern mammal, coupled with this fossil evidence, that clinches the evolutionary progression from reptiles to mammals. In the pouch young of the living opossum, *Didelphis*, which is in essence an embryo that has been born prematurely, we can see exactly the same structural relationships as in *Morganucodon*. The tympanic, malleus, incus, and stapes are in exactly the same position as they are in the jaw of the Jurassic taxa, held together in a linear form by an embryonic structure known as Meckel's cartilage. As development proceeds, the cartilage breaks down and the bones become incorporated into the ear. The evidence is unambiguous; the old angular (reflected lamina) = the tympanic, the quadrate = the incus, and the articular = the malleus. For information sake, although it does not pertain here, the stapes, the third middle ear bone, is homologous to the hyomandibular bone of fishes. So the next time you listen to Mozart or The Rolling Stones, say a brief thank you to a long sequence of vertebrates that made the successful transition from water to land and managed to evolve an exquisite hearing apparatus in the process.

Suggested Readings

Allin, E. F. and J. A. Hopson. 1992. Evolution of the auditory system in Synapsida (mammal-like reptiles and primitive mammals) as seen in the fossil record; pp. 587–614. In D. B. Webster, R. F. Fay, and A. N. Popper (eds.), *The Evolutionary Biology of Hearing.* Springer-Verlag, New York.

Hopson, J. A. 1987. The mammal-like reptiles: a study of transitional fossils. *American Biology Teacher* 49:16–26.

Hopson, J. A. 1994. Synapsid evolution and the radiation of non-Eutherian mammals; pp. 190–219. In Spencer, R. S. (ed.), *Major Features of Vertebrate Evolution, Short Courses in Paleontology No. 7.* The Paleontological Society, University of Tennessee, Knoxville.

Hopson, J. A. and H. R. Barghusen. 1986. An analysis of therapsid relationships; pp. 83–106. In N. Hotton III, P. D. MacLean, J. J. Roth, and E. C. Roth (eds.), *The Ecology and Biology of Mammal-like Reptiles.* Smithsonian Institution Press, Washington, DC.

Hopson, J. A. and J. W. Kitching. 2001. A probainognathian cynodont from South Africa and the phylogeny of nonmammalian cynodonts. *Bulletin of the Museum of Comparative Zoology* 156:5–35.

9

Evolving Voles and Muskrats

". . . if everybody on Earth died in a calamity and, a thousand years from now, an alien civilization visited Earth and began a paleontological dig in Russia, their earliest record of the modern muskrat would mark the twentieth century."

Voles, lemmings, and muskrats are common rodents, usually classified in the mammalian Family Arvicolidae. Arvicolids are found in virtually every habitat in the northern hemisphere; even deserts have their arvicolids. Voles and lemmings represent the primary diet for many northern hawks and other carnivores. Although there are certainly differences among them, most voles have short compact bodies, reduced eyes and ears, and grizzled grayish to brownish fur (**Fig. 9.1**). The tail is short in many species. Grazing voles make paths through the grass, known as "runways," that are used by many animals species. They also burrow prodigiously in the ground, as many gardeners will angrily attest. Voles tend to be very aggressive and do not tolerate the presence of related species well. A few are semiaquatic, feeding on soft vegetation along the edge of streams. Lemmings may also be brownish and nondescript, like the bog lemmings, but the more northern tundra lemmings can be quite attractively colored. Voles and lemmings are prone to display extreme population fluctuations, where thousands or sometimes even millions of the little creatures can virtually denude the landscape of forage. These population explosions are usually followed by a rapid decrease or "crash." Carnivore populations often cycle with vole and lemming cycles.

Muskrats are amazing creatures. They are the world's largest arvicolid rodent, with some individuals averaging over 1 kg in weight. They are aquatic, with webbed feet and soft silky fur. Although they often build lodges like their distant rodent cousins the beavers, more often muskrats dig burrows in the banks of lakes or ponds to obtain shelter and bear their young. They mostly eat freshwater bivalves and aquatic vegetation. Although there may

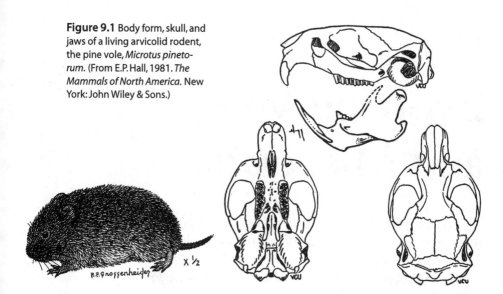

Figure 9.1 Body form, skull, and jaws of a living arvicolid rodent, the pine vole, *Microtus pinetorum*. (From E.P. Hall, 1981. *The Mammals of North America*. New York: John Wiley & Sons.)

be some extinct relatives in the Old World, the only living muskrats were in North America until recently. They were introduced into Europe earlier this century and have now spread virtually throughout Eurasia. My Russian friend and colleague Alexey Tesakov tells me that the muskrat has become an index fossil for the modern period in Russia. That is, if everybody on Earth died in a calamity and a thousand years from now an alien civilization visited Earth and began a paleontological dig in Russia, their earliest record of the modern muskrat would mark the twentieth century. This is an amazing testimony to how fast a mammal species can disperse, particularly one that is aquatic, and it supports a theory I have had for some time that aquatic mammals do not speciate very frequently, because rather than get isolated in separate tributaries they freely navigate between them, exchanging genes. As discussed in Chapter 2, it is generally assumed that in order for new species to arise there must have been a geographical separation between ancestral and descendant populations for some time.

Luckily for us, most modern arvicolids can be differentiated on the basis of their dentition, and this is the key to following evolutionary transitions among them. It is a plain fact of life that most mammalian paleontologists have to be part dentist. Actually, they probably know more about dental development and anatomy than do dentists, the latter who generally restrict their studies to one species, us. The identification and classification of fossil arvicolid species is determined virtually exclusively on the basis of dental anatomy. Before we begin to review dental evolution in the arvicolids, some basic anatomy and vocabulary should be introduced.

All mammals have two sets of teeth, milk (or deciduous) and permanent teeth. Many groups also have four types of teeth; from front to back on one side of the palate or lower jaw these are incisors, canine, premolars, and molars. Although there is always only one canine, the number of other kinds of teeth varies considerably between different mammal groups. The incisors, canine, and premolars have deciduous teeth, but the molars do not. All species in the mammalian order Rodentia lack canines and have a total of two upper and two lower incisors. These teeth have no roots and are therefore ever growing. It is essential that the upper and lower incisors occlude (meet) properly, and if they do not some weird things can happen. I once found a woodchuck skull in which one of the lower incisors had grown up and over the skull, to eventually curve back and penetrate through the skull roof. I do not know if that eventually killed the beast, but it is amazing that the animal was able to survive to that point.

Arvicolid rodents have three molars in both the upper and lower jaws. The space left by the absence of the canine and premolars is called a diastema. All arvicolids have similar flat-crowned teeth that seem to be well adapted for grazing on grasses and sedges, though a few species that live in the forests have a more mixed diet. The occlusal surface of arvicolid molars appears rather geometric in design, and it is both the two-dimensional flatness and simple almost linear geometry that allows paleontologists to make careful measurements and study of their patterns and to discriminate between species. From occlusal view, the pattern of each molar appears to be composed of a series of triangles, which is what the segments are formally called. Each triangle is made up of a tough outer enamel edge and a softer dentine core (our teeth have enamel over a dentine core, but the dentine is not normally exposed to wear). Although other teeth are occasionally useful for identification, arvicolid specialists rely almost exclusively on the first lower molar, or m1, and the third upper molar, or M3, for their work. This is for a very simple reason. These are the most variable teeth, and when new species appear, the m1 and M3 generally change in a recognizable and consistent manner, whereas the other molars often do not change. As we shall see, these teeth may also change within the same species through time.

The earliest arvicolids evolved in Eurasia from another group of rodents, the cricetine murids. The Muridae include most other "rat-like" rodents, including the Norway rat, house mouse, and many other kinds of small mice throughout the world. The cricetines are a subset of the Muridae that include such animals in the United States as cotton rats, white-footed mice, deer mice, harvest mice, and the pack rats. Some paleontologists prefer to view the cricetines in their own family, the Cricetidae. The earliest voles likely ancestral to modern species are recorded from deposits of late Miocene Age in Europe and North America and are all referred to the genus *Promimomys*

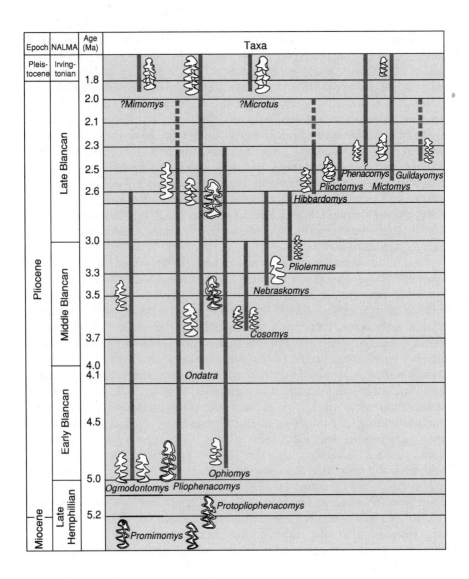

Figure 9.2 Geological record of North American arvicolid rodents. Examples of the chewing surface (occlusal) view of the first lower molar (m1) provided. General evolutionary trends in many lineages include: 1. increase in size, 2. increase in height of crown (hypsodonty), 3. increase in complexity (addition of triangles to occlusal surface), 4. addition of cement, and 5. loss of roots. Illustrations not to same scale; *Ondatra*, the muskrat, is much larger than the others. *Microtus*, a genus with rootless, ever-growing molars, is the most successful of the modern arvicolids, with more than 100 species worldwide.

(**Fig. 9.2**). The first lower molar of *Promimomys* is simple, and the four main areas of the tooth are considered directly homologous (= structurally the same as) to the same areas of a cricetine m1. As in all cricetines, these early arvicolids had roots on the teeth that fit in sockets, called alveoli (alveolus is singular), in the lower jaw. It is important to note at this point that the root of a rodent molar develops embryonically separately from its crown. They fuse early during development, but often after birth.

During the Pliocene and Pleistocene, voles and lemmings explosively radiated worldwide into hundreds of species. Because they diversified so quickly and because they are often ubiquitous in fossil assemblages, they make ideal subjects for studies of evolutionary change. The genus *Mimomys* was especially common in the Old World throughout Pliocene and Pleistocene time (the last five million years), and at least two species of *Mimomys* and a number of close relatives are also known from the New World. Toward the later part of the Pliocene, about 2 million years ago (Ma), *Mimomys* species with extraordinarily high-crowned molars began to appear in European and Russian fossil localities. Studies by Laurent Viriot (U. Burgogne) have shown that in these advanced species the roots fused with the crown later in life than they did in ancestral species, because many molars recovered lack roots at fairly advanced developmental stages. Sometime after 2.0 Ma the first arvicolid molars are recovered that lack roots entirely. Then, almost instantaneously from the standpoint of geological time, species with these kind of teeth are found throughout the Holarctic region and eventually replace almost all the species with rooted molars. In the New World alone more than 20 species are recognized, and these are the common voles of our fields and prairies. Species with rooted molars are now restricted to forests and aquatic ecosystems. It may be that these were the original habitats for arvicolids with rooted molars, but it may also be that the species with rootless molars were more aggressive, with higher reproductive rates, and just crowded out the less advanced species in numerous competitive encounters, restricting the surviving species to less optimum habitats. Rootless arvicolid molars are also ever growing, like rodent incisors. That is, as the crown is worn down by chewing abrasive food, it is replaced by new growth from the bottom. In rooted molars, growth ceases after the root and crown fuse, and as the crown wears down it is not replaced. Conceivably, this could give voles with rootless molars a decided advantage if the species were long lived. However, in nature these animals rarely seem to live more than a year, and few living species with rooted molars are found with their teeth worn down to the nub, so the selective advantage of this modification remains uncertain. The change in molar development may instead be correlated with overall hormonal changes during development that also resulted in higher reproductive rates.

The fossil history of arvicolid rodents provides numerous examples of changes that occur both in lineages (within the same species) and between somewhat more distantly related groups, such as genera (e.g., between *Promimomys* and *Mimomys* or *Pliophenacomys*). Remember that brief periods of evolution within lineages is considered microevolution, whereas more dramatic evolutionary modifications, which may take hundreds of thousands or millions of years to play out, are termed macroevolution. Changes in morphology within living species, for instance, qualify as microevolution, whereas the large-scale patterns expressing the evolution of horses or some other clade throughout the Cenozoic are examples of macroevolution. As we shall see with the muskrat fossil record, sometimes the distinction is not so obvious, as some fairly pronounced change can occur within the history of a single species.

North American Voles

Loss of Molar Roots
Almost everyone living at northern latitudes has seen a *Microtus* vole of one sort or another. If you own cats, then you have certainly seen them, likely at your doorstep. These ubiquitous rodents have been in North America at least since the beginning of the Pleistocene Period (the Ice Ages) at about 1.8 Ma. Their earliest immigration to the New World is documented by their appearance in the Nash 72 fossil quarry in Meade County, Kansas. This quarry lies 3.5m above a volcanic ash in the same sequence that has been radiometrically dated at 2.10 Ma. Although members of the genus *Microtus* may have arrived in the New World at an earlier time, the Nash 72 occurrence marks their first appearance on the Great Plains of North America and is the only securely dated early record on the North American continent. *Microtus*, possessing the rootless molars mentioned above, did not evolve in North America. Instead, this genus evolved in Eurasia from a species of the extinct genus *Mimomys*, which did have roots on its teeth. Intermediates are known. The most important point for our consideration here is that nowhere in the world do species of arvicolids with rootless molars precede those with rooted molars. The fossil record in North America as well as Eurasia is quite clear; arvicolids with rooted molars appeared first (late Miocene through late Pliocene), followed by a period of time in which there was some overlap with descendants with rootless molars (latest Pliocene through middle Pleistocene). Ancestral arvicolids with rooted molars were almost all gone by the early Pleistocene in North America, but persisted in Eurasia well into the middle Pleistocene. In North America there were many genera of ancient arvicolids with rooted molars, such as *Promimomys, Ophiomys, Ogmodontomys,*

Nebraskomys, Pliophenacomys, Pliolemmus, and *Hibbardomys.* All went extinct on this continent by the beginning of the Pleistocene. In North America, there are only three living taxa with rooted molars: *Clethrionomys* (red-backed voles), *Phenacomys* (tundra voles), and the muskrat, *Ondatra zibethicus. Clethrionomys* and *Phenacomys* species are forest dwellers, and the muskrat is predominantly aquatic. It is probably not accidental that these arvicolids, descended directly from ancient arvicolid groups, are limited to these habitats. *Microtus,* today with many species distributed from higher elevations in Mexico to the Arctic, is the dominant arvicolid in North American grasslands. This is almost certainly a result of their highly efficient foraging behavior coupled with ever-growing molars, high reproductive rates, and aggressive nature.

Increase in Triangles of Meadow Vole Molars

In addition to changes in the root configuration, the crown surface pattern also undergoes considerable modification with time in many arvicolid lineages and species groups. Leo C. Davis made an important contribution to evolutionary biology when he reported the change in triangle numbers on meadow vole dentitions from the Great Plains of the United States through part of the Pleistocene period. This is another wonderful example of phyletic change, which is viewed by some paleontologists, as we have seen, as long-term change within the same species, a form of microevolution. The ancestor for voles of the genus *Microtus* is believed to be a European species named *Microtus deucalion.* This species possessed a simple first lower molar, with only three closed triangles. Later species, such as the meadow vole, *M. pennsylvanicus,* have more triangles on this tooth (**Fig. 9.3**). Meadow vole molars usually have either five or six closed triangles. Occasionally, a few molars with seven triangles are observed. The proportion of five or six triangles in a meadow vole population varies from place to place in North America. The highest proportions of populations with six triangles are found on the Great Plains and diminish as one moves either north or east into fields found within more forested habitats. Davis speculated that this difference may be related to vole diet, but there is no clear proof of that yet. The important point for our consideration here is that there is a well-documented progression from an ancestral five triangle condition about 300,000 years ago to the six triangle form now found on the Plains.

Evolution of Crown Patterns in Prairie Voles

Another phyletic lineage where change in crown pattern can be seen is the progression leading to the modern prairie vole, *Microtus ochrogaster,* from ancestral populations on the Great Plains known as *Microtus pliocaenicus.* The author and other investigators have recovered the remains of *Microtus*

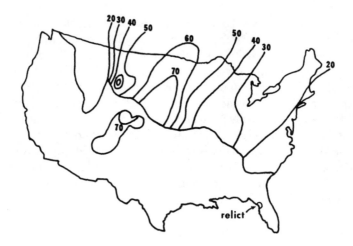

Figure 9.3 Contour map showing the distribution of complexity in first lower molars (m1s) in the living meadow vole, *Microtus pennsylvanicus*. The numbers refer to the percentage of the population with six closed triangles on m1. The remainder of the population usually has five triangles. Davis (1987) demonstrated an increase in percentage of six-triangle forms over the past 250,000 years. (From R.A. Martin, and A.D. Barnosky, eds. 1993. *Morphological Change in Quaternary Mammals of North America*. New York: Cambridge University Press. p. 238.)

pliocaenicus from a number of localities on the Great Plains, 1.5 to 2.0 Ma (e.g., Java, Wathena, Sappa, Kentuck, Nash 72). The first lower molar of *M. pliocaenicus* possesses a posterior loop, three closed triangles, and a simple anterior cap. Populations with slightly more complex molars are known from the younger Conard Fissure site of Arkansas and the Cumberland Cave and Trout Cave No. 2 localities in the eastern United States. These fossils show dental characteristics that lead inexorably toward the molars of the modern prairie vole, which contain five triangles plus an anterior cap. Fossil specimens inseparable from the prairie vole are also recorded from the same depositional areas but at later times. The earliest fossil molars with completely modern prairie vole aspect are probably those from the Cudahy locality of Kansas, securely dated at about 0.67 Ma. The ancestral m1 morphotype still occasionally crops up in living prairie vole populations, but it is very rare.

European Water Voles

The genus *Arvicola* contains two living species, *A. terrestris* and *A. sapidus*, that are usually found in and around streams, lakes, and rivers in Europe. They have rootless ever-growing molars. As discussed in a section above, arvicolids with rooted cheek teeth preceded those without roots in the global fossil record. (Another example of Dollo's Law, which is the general proposition that evolution is irreversible. No arvicolids with rootless molars have been observed to give rise to ones with roots on the teeth.) Another set of evolutionary trends in the molars of arvicolids, which we will also see with the muskrat record in the last section of this chapter, includes increase in height of the crown (hypsodonty); addition of cementum in the infoldings (reentrant angles) between the triangles, presumably to provide strength; and increase in enamel-free areas on the sides of the molars as they increased in height. When viewed from the side of the molar, these enamel-free areas look like elongated extensions of the root up the side of the crown and are therefore called "dentine tracts." Special connective tissue, Sharpey's fibers, attach from the side of the alveolar wall to these tracts, thus helping to anchor the teeth in the mandible (jawbone). From the side of the tooth, changes in the height of the dentine tracts results in changes of the shape of a line marking the enamel–root junction. The German paleontologist Gernot Rabeder has named this line the *linea sinuosa,* and its form is very important in identifying changes within lineages and in separating one arvicolid lineage from another.

Another change seen in molars within the water vole lineage, one that is also seen in other lineages, is a change in relative thickness of enamel on the anterior versus the posterior edge of the triangles. Concentrating our focus only on the lower molars, in ancestral species of *Mimomys* the posterior edges of the triangles are always distinctly thicker than the anterior edges, a condition known as negative enamel differentiation. This is also true of the earliest fossil populations of *Arvicola terrestris,* which can be recognized as the intermediate chronomorph *Arvicolia terrestris /cantiana.* However, modern populations of *A. terrestris /terrestris* display the opposite condition, where the posterior edges of the triangles are thinner than the anterior edges, called positive differentiation. Without going into considerable technical detail, what happens here is that one of the ancestral layers of the posterior edge is lost, which increases the sharpness of this edge, likely as diets of water voles changed through time. Studies by many paleontologists, including Wolf-Dieter Heinrich, Jean Chaline, Laurent Viriot, Thijs von Kolfschoten, and Leonid Rekovets, have recently shown that the oldest ancestral populations in the water vole lineage all possess negative differentiation, roots on the molars, low dentine tracts, and relatively low crowns. Fossils intermediate in age between the oldest specimens of *Mimomys savini* and modern *Arvicola terrestris /terrestris* are intermediate in dental morphology, and the modern populations are dominated by molars with positive enamel differentiation, no roots, high dentine tracts, and high crowns (**Fig. 9.4**).

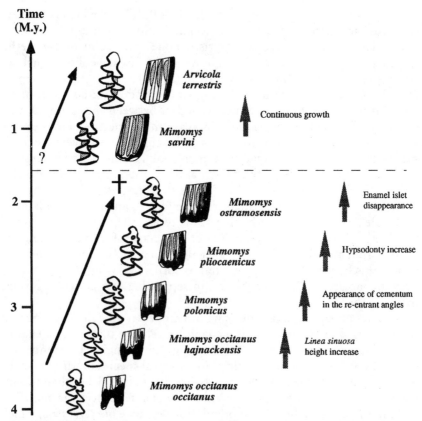

Figure 9.4 Dental evolution in two arvicolid rodent lineages from Europe. Both represent examples of phyletic change, and the constituent populations would not be named this way by the author of this book. Rather, the intermediate stages would be different chronomorphs of only two species, one below and one above the line. Nevertheless, the anatomical changes are well documented and repeated in several different arvicolid lineages. Note especially the eventual loss of roots, resulting in an ever-growing condition that replenishes the chewing surface. (From D. Neradeau, L. Viriot, J. Chaline, B. Laurin, and T. van Kolfschoten. 1995. Discontinuity in the Plio-Pleistocene water vole lineage. *Paleontology* 38:78.)

The Muskrat Ramble

The North American muskrat, *Ondatra zibethicus,* is a textbook example of phyletic evolution. The ancestor to this species probably dispersed to North America across the Bering Land Bridge connecting Siberia with Alaska about 4.5 to 5.0 Ma. The muskrat then spread quickly throughout North America, and fossil remains are known from about 4.2 million years to the present time. It is one of the best known and complete evolutionary histories of any mammal species. As with the other arvicolid species considered previously, the trends that characterize muskrat evolution are mostly confined to the dentition. But it is amazing what can be done with teeth, and we can say a lot about the history of muskrats. One of the keys to this is being able to accurately estimate body weight (or, more properly, mass) in extinct animals, and a bit of a digression about this is appropriate here.

It turns out that body mass in grams or kilograms is correlated very strongly with various bone and tooth sizes in most animals (**Fig. 9.5**). That is, as an animal's skeletal and dental parts increase in size, so does their mass. Scientists then can take the masses of modern animals and plot them against some bone or tooth measurements on a graph and generate a mathematical formula that can accurately estimate the mass of an animal from just the measurements, for example:

$$W = 0.71L^{3.59} \tag{1}$$

where W is the mass in grams and L is the length of the first lower molar in millimeters. (If you are wondering why we do not use "M" instead of "W," it is because "M" has widespread acceptance as the abbreviation for metabolism, or metabolic rate. "W" was used originally for weight, and after scientists began converting to the metric system, "W" just stuck for mass.) This equation was calculated from hundreds of m1 measurements and body masses in living arvicolid rodent specimens from various museums. From this equation we can accurately estimate the average weights of the various muskrats from 4.2 Ma until today. And there is more we can get from this procedure, because it turns out that many biological variables of an animal are also correlated with body mass, such as metabolic rate, lifespan, number of offspring, and home range (the area in which an animal wanders to satisfy its biological needs). Consequently, if we can accurately estimate body mass in an extinct animal, we can also get a reasonable approximation of these attributes. We will get back to this later.

Phyletic change is evolutionary change without branching, without the appearance of a new descendant species that is contemporaneous with its ancestor. We saw in a previous chapter that some scientists prefer to arbitrarily name species in a phyletic sequence, but we also examined this notion and saw that it was flawed for a variety of reasons. There has been only one member of the genus *Ondatra* alive at any point in its more than four million year history in North America. It has changed considerably since that time, but there is no evidence for two or more contemporaneous species. We are blessed with almost a complete record for this species, *Ondatra zibethicus,* with hundreds of fossil specimens represented in many collections, and we can see how evolution took place in all the links of the chain connecting the earliest ancestor with its latest living descendant. In 1985 Leonard Krishtalka and Richard Stucky presented an informal classification system in which inter-

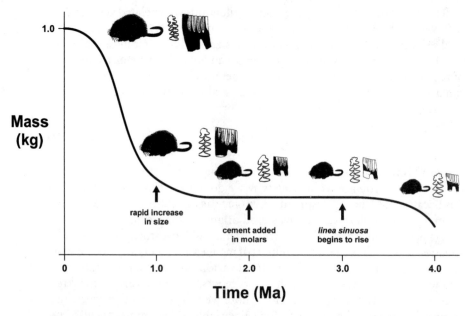

Figure 9.5 Evolution of size and molar anatomy in the North American muskrat, *Ondatra zibethicus*. The history of our muskrat is one of the best-documented cases of phyletic change in the mammalian fossil record. Speciation may have been minimized in this animal because of its aquatic lifestyle, with streams and rivers acting as genetic conduits. (Illustration prepared by Carl Woods, Murray State University, Murray, Kentucky.)

mediate populations in a fossil phyletic sequence were called "lineage segments." The benefit of this approach is that it better represents biological reality and begins to restrict the number of species in the fossil record to those that arose by cladogenesis, or lineage branching, which is the only biological process that produces real species diversity. However, the lineage segment concept is too restrictive. Morphological evolution within species appears to be more complicated than a simple linear response. Mosaic evolution (discussed in Chapter 2) seems to be more common, in which characteristics may change in some populations but not in others. From the basis of the fossil record, this means that populations (samples) with slightly more advanced features may be found that were contemporaneous with those that were not so advanced. The lineage segment concept does not reflect this dynamic very well, and in 1993 I introduced the term *chronomorph* to deal both with fossil samples that might be lineage segments or might also overlap in time. Chronomorphs remain as informal units, with names that often reflect species that had been previously named in the literature as full species. For example, in 1941 the late Claude W. Hibbard named an extinct muskrat from southwestern Kansas as *Pliopotamys meadensis*. The chronomorph version of this species is *Ondatra zibethicus /meadensis*. The first two names tell us the genus and species and the third name tells us the chronomorph. All modern muskrats are included in the chronomorph *Ondatra zibethicus /zibethicus*.

Five chronomorphs of the muskrat are recognized, each characterized by a set of features and a general size range. These characters are all dental, primarily of the first lower molar. In general, muskrat molars became larger and higher crowned, in addition adding cementum between the triangles. These changes are related to overall increase in muskrat body size and the associated need to increase the dental area to process more food. An increase in crown height also theoretically increases the life span of an individual and therefore its reproductive potential. Animals with shorter crowns would have died young, perhaps without reproducing or without leaving as many offspring as those with higher crowns. Just in case there are any doubts that this can be an important factor, my colleague Larry Martin at the University of Kansas noted in a recent article that farmers fit their cows with artificial dentures to increase their productive life span. We will see this trend again within horse evolution. It is a commonly encountered phenomenon in mammalian herbivore evolution, probably because it is so powerfully advantageous; that is, it so often enhances fitness within a species. However, unlike the much smaller advanced terrestrial grazing arvicolids we saw earlier, *Ondatra* teeth never became ever growing; roots are always present.

Cementum is a hard substance found in the dentition of many mammalian groups with high-crowned, or hypsodont, molars. It appears to help solidify the tooth against the many abrasive forces associated with chewing. Cementum is not found in the earliest arvicold rodent ancestors and appears within certain lineages and not others. We have no experimental evidence to explain this observation, though it seems likely that most arvicold groups possess the gene to produce cementum. Among the modern arvicolids, cementum is present in all the advanced terrestrial and aquatic grazers, but not in those species that appear to be relics of early arvicolid evolution that are generally found in forested habitats or are not in competition with advanced species of the genus *Microtus*. Cementum is first seen in the muskrat lineage with the */idahoensis* chronomorph, about 2.6 Ma, but the cementum is poorly developed and, at least in the few molars available for study, does not completely fill the area between the triangles. It becomes well-developed subsequent to about 2.5 Ma (see Fig. 9.5).

Triangles were added to the muskrat m1 as it increased in size through time, generally early during development. The number of triangles of the ancestral juvenile chronomorph mimics the number in its direct adult descendant chronomorph. Juvenile teeth are always more complicated than adult teeth in muskrats because the pattern tends to simplify as it wears down. The size of the first lower molar more than doubled over the course of about 3.75 million years, from about 4.1 mm to more than 8.0 mm, but this size increase was not gradual. Most of it was confined to the last 670,000 years. Also, when we apply Eq. (1) above to these molar dimensions, we discover that muskrat body size much more than doubled during this period of time. This is because of basic physical laws that relate mass and linear measurements in any similar-shaped object as it increases in size and can be seen in the power of 3.59 in Eq. (1). If body mass were to increase at the same rate as length of the tooth, then this exponent would be 1.0, and a doubling of tooth length would, indeed, result in a doubling of mass. But the relationship of molar length to mass in arvicolid rodents follows a general physical law for all objects in the known universe that, stated simply, is as follows: As an object increases in size, its mass increases at a much faster rate, usually to the cube power, relative to any linear dimension taken on the same object. So, a muskrat with an average m1 length of 4.14 mm has an average mass of 0.12 kg, whereas one with an average m1 length of, say 8.3 mm, has an average mass of 1.42 kg, an increase of more than 10 times. This increase in size has a profound affect on an animal's biology. Compare the values for a variety of biological attributes in **Table 9.1** for an ancestral small muskrat and a later large one. Small muskrats foraged over small areas and had brief life spans, high metabolic rates, and large population sizes. Modern muskrats have much lower mass-specific metabolic rates, longer life spans, small population sizes, and forage over much larger areas. It may seem like magic or fantasy to make these kinds of pronouncements about animals dead for millions of years, but the mathematics cannot lie, and these relationships hold within all mammal groups. The mathematical exponent in Eq. (1) may differ from group to group, but the basic relationship between size and all of the variables in Table 9.1 remains the same. (Mass is, in my opinion, the single most important control variable in the known universe. Of course, the expression of these physiological and life history traits must be under hereditary control, but the stability of the relationship between mass and these traits among all organisms argues for an influence of mass on hereditary expression, if for no other reason than it has such a profound influence through the related force of gravity.)

One of my former students, Richard Tedesco, and I showed many years ago that muskrat molars change shape as they evolve, and the method by which they do this can be viewed as a reflection of the way shape changes might come about in other body regions in muskrats and other animals. The width of ancestral muskrat teeth increased at a faster rate as they grew than

Table 9.1 Comparison of important bilogical variables in ancient (*Ondatra zibethicus/minor*) and modern (*Ondatra zibethicus/zibethicus*) populations of the muskrat.

Variable	/minor	/zibethicus
W (kg)	0.12	0.84
M_b (kcal/day)	14.3	61.4
H (ha)	4.4	14.5
D (#/km²)	780	238
L (yrs)	3.9	5.4
O	4.7	5.6

Basal metabolic rate (Mb), home range size (H) in hectares, population density (D), lifespan (L), and number of offspring per litter (O) are estimated from mathematical correlations between body mass (W) and these variables in living mammals. Body mass can be accurately estimated in extinct mammals because of the high correlation between tooth size and body size in living mammals.

they do in modern muskrats. That is, after a certain size had been reached in muskrat evolution within the last million years, the teeth could continue to grow in length at the same rate, but not in width. So large modern muskrats have larger but relatively narrower teeth than their ancestors. Most organic structures are determined by growth vectors of this sort, and it is not difficult to imagine that differences in growth rates, such as expressed by muskrat molars, might govern general shape changes in many organisms. Once again, the fossil record can be seen as an important laboratory within which ideas can be proposed and tested.

Suggested Readings

Barnosky, A. D. 1990. Evolution of dental traits since latest Pleistocene in meadow voles (*Microtus pennsylvanicus*) from Virginia. *Paleobiology* 16:370–383.

Barnosky, A. D. 1993. Mosaic evolution at the population level in *Microtus pennsylvanicus*; pp. 24–59. In R. A. Martin and A. D. Barnosky (eds.), *Morphological Change in Quaternary Mammals of North America*. Cambridge University Press, New York.

Chaline, J. and B. Laurin 1986. Phyletic gradualism in a European Plio-Pleistocene *Mimomys* lineage. *Paleobiology* 12:203–216.

Heinrich, W.-D. 1987. Some aspects of the evolution and biostratigraphy of Arvicola (Mammalia, Rodentia) in the central European Pleistocene; pp. 165–182. In O. Fejfar and W.-D. Heinrich (eds.), *International Symposium Evolution, Phylogeny and Biostratigraphy of Arvicolids*. Geological Survey, Prague, Czechoslovakia.

Hinton, M A. C. 1929. *Monograph of the Voles and Lemmings (Microtinae) Living and Extinct.* Vol. 1. British Museum (Natural History), London.

Martin, L. D. 1979. The biostratigraphy of arvicoline rodents in North America. *Transactions of the Nebraska Academy of Sciences* 7:91–100.

Martin, R. A. 1993. Patterns of variation and speciation in Quaternary rodents; pp. 226–280. In R. A. Martin and A. D. Barnosky (eds.), *Morphological Change in Quaternary Mammals of North America.* Cambridge University Press, New York.

Martin, R. A. 1995. A new species of *Microtus (Pedomys)* from the southern United States, with comments on the taxonomy and early evolution of *Pedomys* and *Pitymys* in North America. *Journal of Vertebrate Paleontology* 15:171–186.

Martin, R. A. 1996. Dental evolution and size change in the North American muskrat: classification and tempo of a presumed phyletic sequence; pp. 431–457. In K. M. Stewart and K. L. Seymour (eds.), *Palaeoecology and Palaeoenvironments of Late Cenozoic Mammals.* University of Toronto Press, Canada.

Martin, R. A. 2003. Arvicolidae. In C. Janis, G. Gunnell, and L. Jacobs (eds.). *Evolution of Tertiary Mammals of North America.* Vol. II. Cambridge University Press, New York (in press).

Martin, R. A. and R. H. Prince. 1990. Variation and evolutionary trends in the dentition of *Microtus pennsylvanicus* from three levels in Bell Cave, Alabama. *Historical Biology* 4:117–129.

Neraudeau, N., L. Viriot, J. Chaline, B. Laurin, and T. Van Kolfschoten. 1995. Discontinuity in the Plio-Pleistocene Eurasian water vole lineage. *Paleontology* 38:77–85.

Repenning, C. A. 1968. Mandibular musculature and the origin of the subfamily Arvicolinae (Rodentia). *Acta Zoologica Cracoviensia* 13:29–72.

van der Meulen, A. J. 1978. *Microtus* and *Pitymys* from Cumberland Cave, Maryland, with a comparison of some New and Old World species. *Annals Carnegie Museum of Natural History, Pittsburgh* 47:101–145.

10

Fishes with Fingers?

It was ". . . the most beautiful fish I had ever seen. . ."
—Marjorie Courtenay-Latimer

On December 23, 1938, then 32 year old Marjorie Courtenay-Latimer, curator of a museum in the port town of East London, South Africa, had no idea what to expect when she was informed that Captain Hendrick Goosen of the ocean trawler *Nerine* had recently arrived at the dock after a voyage off the mouth of the Chalumna River in the Indian Ocean. Captain Goosen had agreed to inform Courtenay-Latimer when he returned from an ocean stint so she could look over the catch and purchase anything that appeared scientifically interesting. She took a taxi to the dock and examined the fish on deck, finally noticing a strange blue fin poking out from under a bunch of unremarkable fish. After moving the other fish out of the way, she was struck by the beauty of a large fish, ". . . mauve blue with iridescent silver markings. . . ," that she could not identify. Courtenay-Latimer had the unhappy taxi driver haul it back to the museum, where she preserved it as best she could. Without proper refrigeration facilities the innards were eventually discarded and could not later be recovered.

Although one of the administrators at Courtney-Latimer's museum dismissed the new find as nothing more than a "rock cod," she was not so sure. So she sent a description of the fish to fellow scientist J. L. B. Smith of Rhodes University in Grahamstown, about 80 km south of East London. Smith was a chemistry teacher who had developed a fascination with ichthyology (the study of fish). He and his wife would later write the definitive treatise on Indian Ocean fishes. As the story goes, Smith was fairly certain the East London fish was special, but he wanted to do further literature research before he decided to see the thing first hand. He finally arrived in East London in mid-February. According to his own writings, he was visibly shaken because his observations confirmed his suspicion that the fish was a member of a group that had supposedly become extinct more than 65 million years ago (**Fig. 10.1**). The group was then called the Crossopterygii, and the kind of animal represented by Goosen's catch was a coelacanth, which means "hollow

191

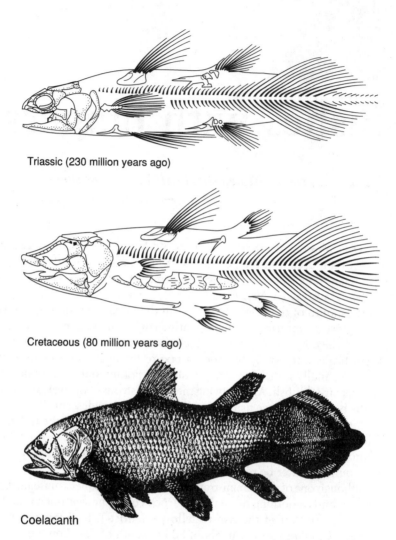

Triassic (230 million years ago)

Cretaceous (80 million years ago)

Coelacanth

Figure 10.1 Fossil and modern lobe-finned fishes. This lobe-finned group, the Actinistia, was believed to have been extinct for more than 65 million years until the coelacanth was discovered by Majorie Courtenay-Latimer in 1938 (see text for details). (top and middle from P.L. Forey. 1988. Golden jubilee for coelacanth *Latimeria chalumnae*. *Nature* 336: 727–732. copyright 1988 Nature Publishing Group.)

spine," named for the structures that support the fins. To a fish expert like Smith, the appearance of a living coelacanth was the exact miraculous equivalent of a dinosaur carcass appearing before a herpetologist.

Smith soon named the remarkable find *Latimeria chalumnae* in a scientific publication in honor of Courtney-Latimer and the Chaluma River. He recognized how important it would be to have the internal organs preserved. He speculated, correctly as it turns out, that the fish might have drifted south in the Mozambique Channel, a deep-water region between Madagascar and the African coast, so he had posters printed up and distributed throughout the coastal islands of southern Africa.

For more than 10 years there was no word. Then, in 1952, Smith received a message from Captain Eric Hunt, a British trader who worked the coast of Africa near Madagascar. Two native fishermen had caught a second coelacanth on a hand line off the Comoros Islands, controlled at the time by the French. The fishermen considered coelacanths inedible and normally threw them back, but they had recognized the fish from posters Hunt had distributed for Smith. As promised on the posters, a reward of 100 British pounds was paid to the fishermen. Hunt had the specimen salted and later injected with formalin. He then cabled Smith in South Africa. Meanwhile, French authorities heard about the discovery and planned to take possession of the fish if Smith did not quickly come to claim it. A plane was provided for Smith by the South African government, and he arrived in the harbor of Pamanzi in the Comoros and whisked the fish off to South Africa. Apparently annoyed that they did not get in on the publicity surrounding the discovery of a coelacanth in their territory, the French authorities closed the islands to non-French investigators until the 1970s, when the islands became independent. To turn this story into further legend, 4 years after the discovery of the second coelacanth, Eric Hunt's abandoned vessel was found on a reef between the Comoros Islands and Madagascar, and he was never heard from again.

The living coelacanth is included with its close extinct relatives in the Actinistia (= Crossopterygii), a group distinct from the living and extinct lungfishes (Dipnoi). These days most researchers do not believe either group is directly ancestral to terrestrial tetrapods like amphibians, but the groups are nonetheless exciting because they give us an important glimpse into the structure and function of part of the initial radiation that led to fully terrestrial vertebrates (animals with backbones). The remainder of this chapter is dedicated to speculating on the transition from water to land and on the fossil animals that we now consider most likely to represent the anatomical links between aquatic vertebrates and their terrestrial tetrapod descendants. As we shall see, the new evidence suggests that feet with digits (fingers) may have evolved in the water, for something other than bearing weight on land. Just for completeness, another population of coelacanths, representing a new species that has been named *Latimeria menadoensis,* was discovered off the coast of North Sulawesi, Indonesia in 1998, almost 10,000 km from the Comoros Islands. This discovery has almost as interesting a story as the original, including a young scientist on a honeymoon . . . but that is for another time.

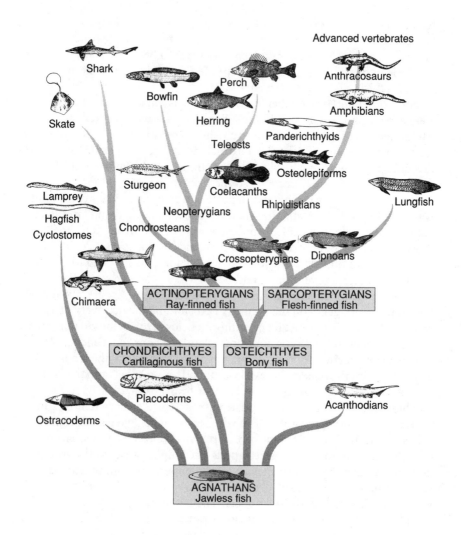

Figure 10.2 A phylogeny of fishes. The earliest fishes, the ostracoderms, had no teeth or jaws and, given their body form, appear to have been bottom dwelling filter feeders. Their external dermal bony armor may have been in response to large aquatic predators called eurypterids, or water scorpions. Neither they nor their presumed living descendants, the hagfishes and lampreys, possessed a bony endoskeleton. Instead, the skeleton was cartilaginous, as it is in sharks and rays. The evolution of jaws, occurring during the Silurian Period, marks one of the great transitions in animal history. From that point onwards vertebrates became the dominant predators on Earth.

What Is a Fish?

Before we get into the fossils, we better make sure we know the difference between a fish and a tetrapod (**Fig. 10.2**). My discussion here is restricted to the bony fishes. Cartilaginous fishes, such as the sharks, skates, rays, and chimaeras, are not related to the lineages of fishes that gave rise to tetrapods.

Most bony fish have internal gills supported by cartilaginous bars for respiration. They also have scales and fins. Their scales have a bony dermal core, which separates them from the scales of reptiles, the latter which are derived entirely from the epidermis (like fingernails). Some fish groups, like the catfish, have secondarily lost scales and are capable of a limited amount of cutaneous (skin) respiration. Amazing as it sounds, lungs (considered as ventral paired outpocketings of the anterior region of the digestive tract, the pharynx) are primitive for fish. They are retained in this position in lungfishes and the African bichir, and although dorsal in position, there is still a connection in gars and the bowfin. All these animals are capable of gulping air to augment respiration with the gills. The most common bony fishes today instead have an air bladder, a single structure that no longer has a connection with the pharynx and is used primarily for buoyancy. The skull in fishes is laterally compressed and composed of many bones, including a set that encloses the gills. The vertebrae of fish are fairly similar and do not have structures known as zygapophyses that are used to connect them one to another. The vertebrae in most fish tend to be biconcave (hollow both anteriorly and posteriorly) and do not, therefore, fit into each other. The ancestral notochord, a cartilaginous rod that characterizes all chordate animals, often runs through the vertebrae as a major support structure, although it is reduced to a series of small pieces between the vertebrae in many advanced bony fishes. The pectoral, or "shoulder," girdle in fishes is composed of a series of bones that articulates with the posterior region of the bones that support the gills. Two bones of the pectoral girdle of advanced bony fishes, the scapula and the coracoid, are considered homologous to the same named bones in the amphibian shoulder. A variety of small bones, known as the radials, supports the fins, which are given shape by a series of either spiny or cartilaginous rays. Ribs are present, but they are fairly slim and do not support the viscera to any extent. A system of small pores that connects to special sensory organs comprises the lateral line system, running from the head along the sides of the body. This is a special adaptation in aquatic organisms for detecting vibrations in water, and holes from this system can be seen in fish skulls. A series of ribs supports the tail (caudal fin), which is the primary propulsive device. Many other features characterize fish (such as a two-chambered heart), but because they do not readily show up in fossil specimens, we will not consider them further here.

Most female bony fish lay eggs that are fertilized externally by males (the *oviparous* condition). A few bony fishes may bear live young that have been developing from eggs retained within the body (the *ovoviviparous* condition), but it is not as common as in the Chondrichthyes (sharks, skates, rays, and chimaeras), where the males of most species have special extensions of the anal fins called claspers that direct sperm into the female's urogenital orifice (cloaca). The eggs of bony fish have no external protection and would soon desiccate if left in the air.

Bony fishes are a diverse lot and include a few primitive groups such as the gars, the bowfin, paddlefishes, sturgeons, the bichir, lungfishes, and the coelacanth, plus the more advanced groups we are most familiar with, such as the tunas, trout and salmon, black basses, perches, eels, cod, minnows, pickerels and pike, flounder and fluke, groupers, and menhaden. Only the lungfishes and coelacanth have further bearing on the tetrapod story. Living lungfishes are freshwater inhabitants found in South America, Africa, and Australia. Like the coelacanth, they are characterized by their lobe-like fins. The African lungfish can withstand extremely dry conditions by burrowing in the mud and remaining in a dormant state, often for years. Lungfishes have long been recognized as generally intermediate structurally between typical fishes and amphibians. The coelacanth is a special animal, as noted earlier in this chapter, primarily because until 1938 it was thought to have been extinct for at least 65 million years.

What Is a Tetrapod?

Why didn't we ask the question "What is an amphibian" instead? In the not too distant past, paleontologists thought that the departure from water and the evolution of a four-footed amphibian body went together (**Fig. 10.3**). After all, they reasoned, the four-footed (tetrapod) condition, complete with digits, must have evolved to support the body on land. All four-footed vertebrates that first came ashore were likely closely related (monophyletic in cladistic parlance) and could therefore be called amphibians. Also, it was generally assumed that all nonamniote terrestrial tetrapods could be considered a clade. (Amniotes are terrestrial vertebrates with an amniote egg. This egg has a series of homologous membranes, such as the amnion and chorion. Amniotes include the reptiles, birds, and mammals.) New evidence complicates the picture considerably. Although it is necessary for successful terrestrial animals to support the body with limbs, it turns out that limbs and digits may have originally evolved first in fully aquatic organisms, sometimes in groups that were not directly ancestral to later terrestrial vertebrates. Also, as Michael Fracasso noted, there were apparently a number of fairly successful nonamniote tetrapod groups that evolved on land that were not especially closely

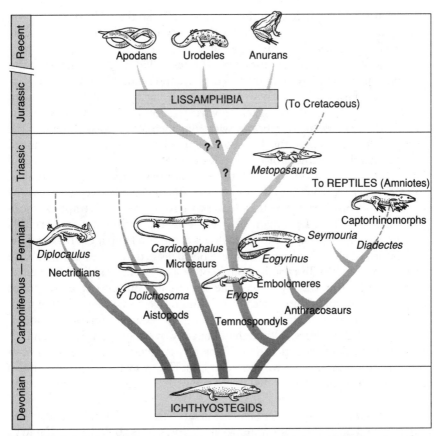

Figure 10.3 A phylogeny for tetrapods. Many early experiments in terrestrialization occurred during the Paleozoic Era after the evolution of feet. The earliest tetrapods were likely all primarily aquatic animals. Two surviving clades of the early tetrapod radiation led to the reptiles (and then to us) and to the living lissamphibia.

related to the groups we label together as the "Amphibia," the latter including the salamanders (Caudata), frogs and toads (Anura), caecilians (Apoda), and the extinct temnospondyls. Instead, there appears to be a cloud of rapidly evolving tetrapod clades during the early Carboniferous, of which the line leading to modern amphibians was merely one. According to the modern rules of classification, if one can recognize a number of such clades (monophyletic groups, each with a separate ancestor and a number of descendant taxa), each must bear a name of equal rank. The Amphibia, therefore, is just one among many. This is a theme we shall see throughout this book: At the start of a successful radiation there are many experiments in structure and ecology, and a mosaic of features often emerges in a number of related groups (see discussion of mosaic evolution in Chapter 2). So the early Tetrapoda (or, colloquially, the tetrapods) includes the Amphibia and other nonamniote tetrapod clades, whether or not they were terrestrial.

The most recent classification of vertebrates includes all lobe-fin fishes and tetrapods in a grand group known as the Sarcopterygii. Actinistians and dipnoans are considered as early experiments among vertebrates that invaded the land but off the direct track toward tetrapods. As noted above, the Tetrapoda includes all vertebrates with four distinct limbs with digits. The group likely ancestral to the Tetrapoda is an advanced group of lobe-finned fishes called the Osteolepiformes, known only as fossils from the Middle and Late Devonian Period. One of these is a fairly famous genus named *Eusthenopteron*, from Late Devonian deposits in Canada (see Fig. 10.4). At least one osteolepiform lobe-fin apparently gave rise to the Tetrapoda, probably during the Late Devonian. (An outside chance exists that the currently accepted group of fossil tetrapods includes evolution from more than one osteolepiform. That is, limbs might have evolved more than once during the Late Devonian. If so, the Tetrapoda would not be a natural group, and further research would be necessary to tease the descendants out into as many groups as there are osteolepiform ancestors.) It should be noted here that the transition from an osteolepiform fish to a tetrapod occurred during a period of around 15 million years. Although not exactly overnight, this is just a minor tick on the clock of geological time. To better appreciate and understand what it means to be a tetrapod, we should at least briefly examine the anatomical features that would be expected to change with the transition from water to land.

There is no more significant change in environment that an organism can experience than going from water to land, unless it is going from a liquid to a solid state (or the reverse directions). Although there are vertebrate animals that do freeze in winter ice and thaw in the spring with no ill effects, in general the solid state does not seem to make for diverse ecosystems. All the senses are affected in the transition to land. The lateral line system, so effective in water to record vibrations, is basically useless on land, because to function the lateral line requires an equal density on both on the inside and outside of the animal. The optical properties of water are different from air, and of course gills are useless for respiration in air. As a much more dense medium than air, water supports the viscera (internal organs), but on land a new anatomical configuration would be necessary both to support the viscera and to allow effective locomotion. Even in the absence of fossils, these considerations would allow an intelligent observer to imagine the construction of animals making such a profound transition. In addition to the development of limbs and digits, we would expect early tetrapods to demonstrate some combination of the following suite of characters:

1. Reduced number of bones in the skull, particularly those associated with the gills;

2. Bony stapes (middle ear bone) derived from the fish hyomandibular cartilage (one of the cartilages that supported the gills);

3. Dorsoventrally flattened skull;

4. Tendency to develop connections between vertebrae;

5. Reduction of notochordal contribution to the backbone;

6. Increased thickness of ribs;

7. Reduction in caudal ribs and tail;

8. Loss of scales.

A much larger list would be presented by a specialist, but this one covers most of the important changes.

Eusthenopteron and *Panderichthys:* Links to Tetrapods

The best example of an ancestral osteolepiform fish is *Eusthenopteron foordi* from freshwater deposits of more than 370 Ma in Quebec, Canada. It is probably one of the most widely studied vertebrates on the planet, along with *Archaeopteryx lithographica* and *Tyrannosaurus rex*. A reconstruction of *Eusthenopteron* forms part of the logo for the Paleontological Society. It is a lobe-fin fish, but it also has a series of features placing it closer to the ancestry of tetrapods than the Dipnoi and Actinistia. Probably the most important is the presence in *Eusthenopteron* of internal nostrils located in the front of the palate, or roof of the mouth. Internal nostrils are absent in the coelacanth but present in all tetrapods. Also called choanae, these nostrils were apparently part of a system allowing water to wash over an olfactory (smell) organ in *Eusthenopteron,* an extension of the brain in the anterior part of the skull. They did not take on a function relating to aerial respiration until at least partial terrestrialization had been accomplished. Osteolepiform fishes were also powerful and adept predators, as indicated by the many sharp teeth, including enlarged "tusks" in the roof of the mouth. According to I. I. Schmalhausen, the premaxillary glands emptying in the mouth at the base of the tusks of *Eusthenopteron* may have produced poison, although in modern terrestrial amphibians these glands mainly secrete mucous. Under high magnification, sections taken from osteolepiform tusks show a special infolded, or "labyrinthine," structure that they also shared with early tetrapods and not with dipnoans and actinistians.

Eusthenopteron was functionally and structurally a fish. It retained a laterally compressed fish-like head, with eyes on the sides of the skull. Pits for the lateral line system in the skull indicate that this system was highly functional. Structures homologous to the main bones of the arm and leg were

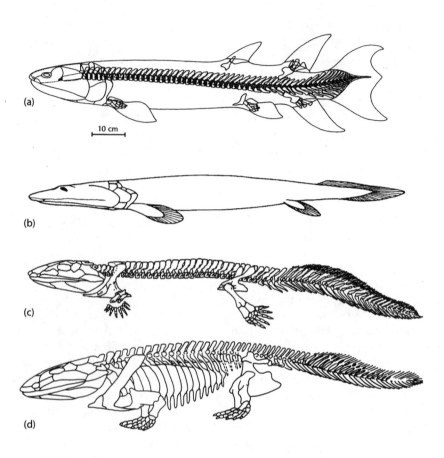

Figure 10.4 Reconstructed body forms Late Devonian vertebrates: the osteolepiform fish *Eusthenopteron* (a), sarcopterygian *Panderichthys* (b), tetrapod *Acanthostega* (c), and tetrapod *Ichthyostega* (d). *Eusthenopteron* and *Panderichthys* were recovered from sediments about five million years older than those that produced the tetrapods. ([a] From R.L. Carroll, 1997. Patterns and Processes of Vertebrate Evolution. Cambridge, UK: Cambridge University Press, p. 300; after S.M Andrews, and T.S. Westoll, 1970. The postcranial skeleton of rhipidipstian fishes excluding Eusthenopteron. *Transactions of the Royal Society of Edinburgh* 68: 391–489. [b] E.I. Vorobyeva, and H.P. Schultze. 1991. Description and systematics of panderichthyid fishes with comments on their relationship to tetrapods. In H.P. Schultze, and L. Trueb. *Origins of the Higher Groups of Tetrapods*. Ithaca, NY: Comstock Publishing Associates. pp. 68–109; [c] M. I. Coates, and J.A. Clack, 1995. Romer's Gap—tetrapod origins and terrestriality. In M. Arsenault, H. Lelièvre, and P. Janvier, eds. *Proceedings of the 7th International Symposium on Early Vertebrates*. Bulletin du Muséum National d'Histoire Naturelle, Paris. pp. 373–388. © Publications Scientifiques du Muséum National d'Histoire Naturelle, Paris.)

present at the base of the fins, but most of the more distal elements of tetrapods, such as carpals, metacarpals, and phalanges (hand and finger bones), had not yet appeared. The full complement of fishy fins was present, with two dorsal fins, pectoral and pelvic fins, and anal and caudal fins. The axis of each fin at this point was basically straight, from the body to the tip. This will have more meaning when we examine the possible genetic/developmental origin of paired limbs.

The next stage in tetrapod evolution is represented by the Panderichthyidae, including *Panderichthys* and *Elpistostege,* Late Devonian aquatic sarcopterygians from Latvia and Canada, respectively (**Fig.10.4**). In *Panderichthys,* for which we have full skeletal material, the two dorsal fins and the anal fin of fishes are gone, leaving only the pectoral and pelvic fins, plus a tail (caudal fin) that is reduced in relative size from that of *Eusthenopteron.* The skull is flatter than in *Eusthenopteron,* a frontal bone is present, and the eyes are more dorsally placed, as in the earliest tetrapods such as *Ichthyostega* and *Acanthostega.* The skull bones also tend to be more consolidated and fused. Yet *Panderichthys* shares a number of fish-like attributes with *Eusthenopteron:* a large number of skull bones, including those of the opercular series that articulate with the pectoral girdle; a well-developed lateral line system; a backbone primarily supported by a cartilaginous notochord; and limbs without hands and feet. The skeletal evidence indicates that *Panderichthys* was exclusively aquatic. Is *Panderichthys* an osteolepiform fish on its way to becoming a tetrapod, or is the combination of skeletal structures it possesses different enough from *Eusthenopteron* that it deserves its own higher taxonomic category? Without feet and digits it is not a tetrapod, but what should we do with an animal like *Panderichthys* that shows such an intermediate constellation of features? Perhaps it is just another missing link.

Experiments with Limbs

A reconstruction of the continents during the Middle and Late Devonian suggests that the ancestral supercontinent Pangaea had begun to split, with North America, Greenland, and Europe (including Russia) separated from the more southern continental group that included South America, Africa, India, Australia, Antarctica, and parts of Asia. The northern continental group is sometimes referred to as Euroamerica. As described by C. R. Scotese and W. S. McKerrow, many of the Devonian fossil sites in Euroamerica with primitive tetrapods probably lay along the equator. It seems a reasonable conjecture that tetrapods arose under tropical or subtropical conditions. Keith Thomson concluded that the earliest lobe-fins diversified in shallow marine environments and only secondarily invaded freshwater habitats. This provides a satisfactory explanation for the distribution of fossil lobe-fins and is also consistent with what we know about living actinitian and dipnoan phys-iology. Because Late Devonian tetrapods are known from both freshwater and shallow marine sediments, as has been suggested by Michael Coates and Jennifer Clack, it is possible that tetrapods evolved independently in both en-vironments.

The Earliest Tetrapods Were Aquatic and Sexy

Sometimes scientific information seems contradictory to common sense. It is not easy to think of the earliest tetrapods as totally aquatic, any more than it is not easy to think of these animals as something other than amphibians. If feet and hands did not develop for walking on land, why did they evolve?

One would not normally think of Greenland as a hotbed for the discov-ery of early tetrapods, but again we need to remember that this island conti-nent lay along the equator during the Late Devonian. Many complete skeletons of the next stage of tetrapod evolution have been unearthed from freshwater deposits in Greenland, from the 1930s until the present day. The work there was begun by G. Säve-Söderbergh and Erik Jarvik and is continued today by Jennifer Clack and Michael Coates. The important taxa recovered are *Ichthyostega stensioei* and *Acanthostega gunnari*. Although found in sediments of the same age, according to Jennifer Clack these species may have lived in somewhat different ecological settings. The best *Ichthyostega* specimens have been found in very fine-grained sediments, suggesting relatively quite waters, whereas *Acanthostega* specimens are often found in sediments deposited in active river channels.

Ichthyostega and *Acanthostega* once again demonstrate the mosaic ex-pression of physical characters that is generated during the early radiation of a successful clade. Both species are advanced over the osteolepiform fishes in possessing hands and feet. This makes them tetrapods. Other features in com-mon with tetrapods include the loss of many typical fish skull bones, a dorsoventrally flattened skull, new attachments called zygapophyses that con-nect adjacent vertebrae, and thicker ribs. However, characteristics that demon-strated an intermediate position between the osteolepiforms (including, for the moment, *Panderichthys*) and later tetrapods include a functional lateral line system, retention of some fish skull bones, lack of distinct occipital condyles on the skull to support the skull on the vertebral column, and large notochordal contribution to the vertebrae.

The differences between *Ichthyostega* and *Acanthostega* probably express a fundamental difference in habitat preference and terrestrialization. *Ichthyostega* appears to have lost internal gills, whereas they were retained in *Acanthostega*. This implies that *Ichthyostega* probably spent more time out of water than did *Acanthostega*. This idea is supported by the unique sturdy overlapping ribs of *Ichthyostega*. Nevertheless, according to Jennifer Clack the terrestrial activity of *Ichthyostega* was probably limited to hauling its hindlimbs around on the shore, much like an elephant seal. Although the hand of *Ichthyostega* is not known, the pectoral girdle is very robust compared with its short and stout hindlimb. The foot of *Ichthyostega* had seven digits.

The limbs of *Acanthostega* were more symmetrically developed than those of *Ichthyostega,* but the presence of internal gills and a longer caudal fin indicates that this animal was primarily aquatic. The paddle-shaped hands and feet each had eight digits. Jennifer Clack recovered the middle ear bone, the stapes, from *Acanthostega* specimens. The stapes has not as yet been positively identified in *Ichthyostega.* This bone in living amphibians connects the tympanic membrane with the cochlea and is responsible for hearing in air. Its auditory function in *Acanthostega* is debatable, but its derivation from the osteolepiform fish hyomandibular bone is not in contention. The stapes, which fits into a concavity called the "otic notch" at the back of the skull in *Acanthostega,* seems also to have formed a solid connection of the palate and the remainder of the skull. A membrane undoubtedly stretched across the otic notch above the stapes, but its function is unclear. It could have been a valve that controlled water flow to the gills, or it could have functioned in underwater hearing. Perhaps it acted in both capacities, not unexpected for a "missing link." Nevertheless, for the middle ear to be fully functional in air, the stapes must be free to vibrate, and this kind of system does not appear in the fossil record until millions of years later, during the Carboniferous.

Getting back to our original problem, of what value would hands and feet be to fully aquatic organisms? Most biologists and paleontologists suggest that the primary use of limbs with digits in the earliest tetrapods is to facilitate movement on the bottom where the animals predominantly feed and live. Let us call this the Bottom Forager hypothesis. We see an analogous structural approach in more advanced species of bony fish that have secondarily adopted a bottom feeding mode, such as the gobies and darters. In these fishes, which do not have the tetrapod arm and leg bones within their fins, we also see the appearance of fan-shaped fins that are used to pull the body along the bottom. In fact, some species of gobies spend considerable time out of water, pulling themselves along the shore with these fins. That may be the story, but I have another hypothesis: that hands and feet made early tetrapods sexier.

Although there are living amphibians that fertilize eggs as they are extruded from the female, without any body contact between the sexes (e.g., *Cryptobranchus* and *Ranodon*), most engage in a reproductive behavior called amplexus, where the male grasps the female tightly. This contact serves to excite the females to lay eggs and apparently may also trigger the final development of the eggs, because amplexus may last for days. This is true whether mating takes place in water or on land. The males of a few salamander species have further devised the trick of internal fertilization, by using either their tail or hindfeet to direct sperm into the female's cloaca (common urogenital opening) before egg laying (e.g., in the wholly aquatic *Amphiuma*). In this scenario, hands and feet evolved to facilitate reproduction, which in turn increased the probability of eggs being fertilized in water. This idea, which we can call the Sexy Tetrapod hypothesis, provides a direct link between the appearance of an evolutionary novelty and species fitness. All we need now is a good idea of how the genetics and development (ontogeny) of amphibians could lead to the appearance of feet and fingers.

A Genetic–Developmental Explanation for the Origin of Tetrapod Limbs

Cell and molecular biology, including genetics, is probably the hottest area in biology these days, and this prominence is likely to continue. Information from these disciplines is leading to many important treatments for human disease, for example. Fundamental knowledge from cellular biological research, though, impacts on all the sciences, including evolutionary biology. We now know that the embryonic development of limbs in all modern tetrapods (amphibians, reptiles, birds, and mammals) is essentially identical. A limb bud forms early, from undifferentiated cells called mesenchyme. These cells proliferate and the bud expands. The expansion is controlled by three genes found in all tetrapods: sonic hedgehog (*shh*), fibroblast growth factor 2 (*fgf2*), and *Wnt7a* (no long name). Each of these genes produces a gradient of molecules (exact nature still undetermined; some may turn on other genes) that influence limb development: *shh* establishes anterior–posterior (front to back) axis of the limb, *fgf2* establishes the proximal–distal (close to the body–away from the body) axis, and *Wnt7a* establishes the dorsal–ventral (up–down) axis.

As noted in Chapter 2, biologists have also discovered a family of genes in animals as diverse as worms, flies, and mice, called the *Hox* genes, that are responsible for the major body segmentation in an anterior-posterior direction (**Fig. 10.5**). This linear *Hox* sequence may also be responsible for chemically conveying information to the developing limb cells, "telling them" where they are along the limb axis. The longer limbs of tetrapods relative to osteolepiform fishes like *Eusthenopteron* may be due, in part, to these *Hox* genes being turned on for a longer time. (Long necks in animals, such as in giraffes and ostriches, for example, may be primarily due to certain *Hox* gene switches left on for a longer time during development.) Paolo Sordino and his colleagues have found that in the development of mouse (a tetrapod) limbs, both the *Hox* genes and *shh* are turned on late in development, but this expression is absent in the fin development of a fish (zebra danio) they studied (**Fig. 10.6**). Their work also confirmed an earlier hypothesis by Neil Shubin and

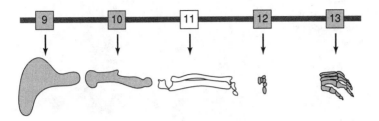

Figure 10.5 Specific *Hox* genes are now known to be responsible for limb formation. This illustration shows the *Hox* clusters responsible for forelimb formation in the laboratory mouse. Mutations that nullify the function of the genes *Hoxa-11* and *Hoxd-11* result in loss of the radius, ulna, and some of the carpals. (From A.P. Davis, D.P. Witte, H.M. Hsieh-Li, S.S. Potter, and M.R. Capecchi, 1995. Absence of radius and ulna in mice lacking *hoxa-11* and *hoxd-11*. *Nature* 375: 791–795. Copyright 1995, Nature Publishing Group.)

Pere Alberch that the primary axis of hand and foot development in tetrapods is curved, as opposed to straight in the development of the osteolepiform fin. Digits are added during development from posterior to anterior (from your pinky to your thumb). Thus, hands and feet are additions, not modifications, to the ancestral lobe fin. Apparently the earliest expression of this new developmental regime led to numerous digits, not merely the four or five we see in later ancestral terrestrial tetrapods.

The Transition to Land

In a recent paper on tetrapod origins, Ted Daeschler and Neil Shubin listed eight Late Devonian genera, including (besides *Ichthyostega* and *Acanthostega*) *Elginerpeton* from Scotland, *Hynerpeton* from the United States (Pennsylvania), *Metaxygnathus* from New South Wales, Australia, *Obruchevichthys* and *Ventastega* from Latvia, and *Tulerpeton* from Russia. Daeschler later described a second tetrapod, *Densignathus,* from Pennsylvania. The material of these taxa is somewhat fragmentary as compared with that of *Ichthyostega* and *Acanthostega* and is not considered here in detail. At least at this point, these taxa do not reveal anything of significance about the fish–tetrapod transition that we have not already covered. Probably the most important aspect of these fossils is that they serve to remind us that early aquatic tetrapods were a diverse and widespread group during the Late Devonian and that we can expect to see a mosaic combination of characteristics in them as they all "experiment" with ways to cope in shallow warm waters and to make the best use of the nearby land surface.

Figure 10.6 Axis of development in vertebrate limbs. In the mouse limb (a, left), the fingers develop in an arc, whereas in the fish limb (a, right), represented by a zebra danio, bony elements form in a linear fashion. Diagrams in (b) show the hypothetical origin of limb structure in a tetrapod (left), Devonian osteolepiform fish *Eusthenopteron* (center), and modern bony fish (right). These studies suggest that fingers in tetrapods were a new addition during the Devonian. (Courtesy of Prof. Denis Duboule, Universite de Geneve, Switzerland.)

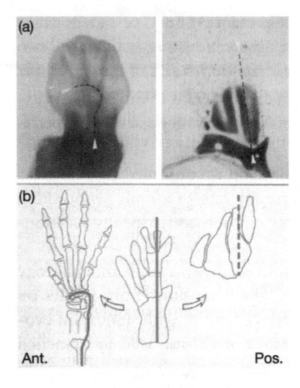

With the recent observations by Jennifer Clack showing that *Acanthostega*, with its four well-developed limbs, was likely fully aquatic, we have a problem in determining which early fossil tetrapods would have been fully terrestrial. If the hands and feet evolved first in water, possibly for reproductive purposes, then we need to find some other skeletal markers indicating terrestrialization. However, this is tricky, because even modern amphibians are, after all, amphibious; they often spend part of their life in water. And it seems pretty clear that most of the later Devonian tetrapods had the anatomical capabilities to (and probably did) spend some time on land, either basking or foraging. One might believe that the presence of an aerial hearing apparatus (composed of a slim stapes free to vibrate against a tympanic membrane) would be the clearest indicator of full terrestrialization, but neither living salamanders nor caecilians have this system, and this absence may very well have been the retention of an ancestral condition. However, the limb and girdle elements are so sturdy in some later Paleozoic stem tetrapods it seems likely that by at least the Late Carboniferous, if not earlier, some had begun to fill the ecological niches now filled by the living amphibians.

The Carboniferous Period (sometimes broken into an earlier Mississippian and a later Pennsylvanian Period, for a total of about 363 to 247 million years ago) was a time of warm and humid tropical and subtropical conditions, where terrestrial habitats were dominated by large ferns and lycopods (fern-like plants called seed ferns) that had the growth habits of trees and bushes. Lands snails appeared during this time, and the forests were populated with an array of insects, crustaceans, and arachnids. The great coal beds of the world formed over millions of years from their dead organic components (this was not the only time coal formed in the past, but it was the most geographically and temporally extensive time of coal formation). Much of our knowledge of Carboniferous tetrapods comes from specimens taken from major coal seams, sometimes from inside fossilized hollow trees. The stem tetrapods may be for convenience lumped into three main groups: the Temnospondyli, including the extinct stereospodyls and the lissamphibia (includes the modern Amphibia), with four digits on the front foot and five on the rear and relatively flat skulls; the Anthracosauria (including the extinct embolomeres and the amniotes), with five digits front and five rear and deep skulls; and the Baphetidae (e.g., *Eucritta*), with strange keyhole-shaped orbits. Most of the temnospondyls, which make up most of the early Carboniferous stem tetrapods, were primarily aquatic, as indicated by the well-developed lateral line canals that appear in many skulls. However, some, such as the famous *Eryops*, had such sturdy limbs and girdles that it is almost certain they could traverse on land regularly. Temnospondyl body forms included salamander through crocodile-sized animals, and all were carnivorous, as indicated by numerous sharp teeth and palatal tusks, or fangs. Although an otic notch is found in all species, the presence of an aerial hearing apparatus has not been confirmed in any.

So what can we conclude about the transition from water to land? Taking liberty with a diverse body of evidence for the purposes of summarizing this chapter, we can say with some assurance that the osteolepiform fishes (e.g., *Eusthenopteron, Panderichthys*) gave rise to the earliest tetrapods, which were primarily aquatic (e.g., *Acanthostega*). The hands and feet may have initially evolved in water, perhaps not primarily as a locomotor device but for males to grasp females (amplexus), thus increasing the probability of fertilizing the eggs. A number of Late Devonian tetrapods appeared, some perhaps with the ability to spend considerable time out of water (e.g., *Dendrerpeton*). By the early Carboniferous (Mississippian), stem tetrapods had radiated into a number of body types, but most remained aquatic. A few groups were probably more terrestrial (e.g., *Eryops*), although it is debatable if they had evolved an aerial hearing system. They were aware of "sound" by detecting vibrations through the ground, as in living salamanders. One of the later

Carboniferous stem tetrapod groups, possibly among the embolomeres, was ancestral to the amniotes (e.g., first reptiles).

Suggested Readings

Ahlberg, P. E. and A. R. Milner. 1994. The origin and early diversification of tetrapods. *Nature* 368:507–514.

Andrews, S. M. and T. S. Westoll. 1970. The postcranial skeleton of *Eusthenopteron foordi* Whiteaves. *Transactions of the Royal Society of Edinburgh* 68:207–329.

Carroll, R. L. 1997. *Patterns and Processes of Vertebrate Evolution.* Cambridge University Press, New York.

Clack, J. A. 1994. Earliest known tetrapod braincase and the evolution of the stapes and fenestra ovalis. *Nature* 352:392–394.

Coates, M. I. and J. A. Clack. 1991. Fish-like gills and breathing in the earliest known tetrapod. *Nature* 352:234–236.

Daeshler, E. B. 2000. Early tetrapod jaws from the Late Devonian of Pennsylvania. *Journal of Paleontology* 74:301–308.

Daeschler, E. B. and N. Shubin. 1995. Tetrapod origins. *Paleobiology* 21:404–409.

Daeschler, E. B., N. H. Shubin, K. S. Thompson, and W. W. Amaral. 1994. A Devonian tetrapod from North America. *Science* 265:639–642.

Genealogy Goosen [website]. 2002. *www.goosen.org/index.html.* Accessed April 17, 2003.

Jarvik, E. 1996. The Devonian tetrapod *Ichthyostega. Fossils and Strata* 40:1–206.

Meyer, A. 1995. Molecular evidence on the origin of tetrapods and the relationships of the coelacanth. *Trends in Ecology and Evolution* 10:111–116.

Noble, G. K. *The Biology of the Amphibia.* Dover, New York.

Schmalhausen, I. I. 1968. *The Origin of Terrestrial Vertebrates.* Academic Press, New York.

Shubin, N. H. 1995. The evolution of paired fins and the origin of tetrapod limbs: phylogenetic and transformational approaches. *Evolutionary Biology* 28:39–86.

Spencer, R. S. (ed.). 1994. *Major Features of Vertebrate Evolution. Short Courses in Paleontology* 7. The Paleontological Society, University of Tennessee, Knoxville.

Thomson, K. S. 1993. The origin of the tetrapods. *American Journal of Science* 293:33–62.

11

A History of Upstanding Primates

"...neanderthals were gone by around 30,000 years B.P., when we see the famous Cro-Magnon human populations in Europe. My bet is that the prime human characteristics of intelligence, cruelty, and greed won the day. Just watch the stock market. ..."

Thomas Henry Huxley was not the first to suggest that humans had a considerable antiquity, but his book, *Zoological Evidences of Man's Place in Nature,* published in 1863 was probably the first treatment of its type to be taken seriously. In this book Huxley illustrated the skeletons of the great apes compared with that of modern humans and concluded that gorillas and humans were more similar to each other than either were to the monkeys (**Fig. 11.1**). From these observations Huxley wisely (and correctly) predicted that eventually prehuman fossils would be discovered that linked apes and humans and that these fossils would likely be discovered in Africa. Charles Darwin also dealt with the problem of human origins in an 1871 publication, *The Descent of Man,* carrying his theory of evolution to its logical conclusion, namely that humans had evolved in an unbroken chain from the first living things on Earth. In one of his many prophetic moments, Darwin also correctly identified Africa as the likely cradle of human civilization:

> It is ... probable that Africa was formerly inhabited by extinct apes closely allied to the gorilla and chimpanzee, and as these two species are now man's nearest allies, it is somewhat more probable that our early progenitors lived on the African continent than elsewhere" *(from* The Descent of Man*).*

It is not difficult to understand the shock these theories produced in their day, and it is no surprise that many other suggestions of associations between

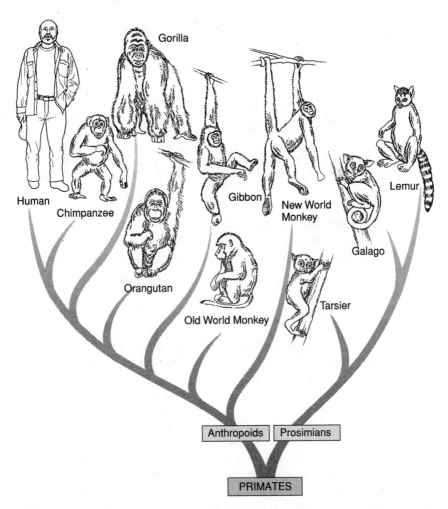

Figure 11.1 A family "tree" (pun intended) for living primates. Humans share over 98 percent of their DNA with living chimpanzees, and fossil evidence suggests the two groups diverged between 7 and 8 Ma.

archaic humans and prehistoric animals had been squelched earlier. Millions of people today live under the mistaken belief that the Earth is young, less than 10,000 years in age, and that humans were created *de novo* by a supreme being. This is essentially the identical position held by the Archbishop James Ussher of Ireland who, in 1650, proclaimed that the Earth was created in the year 4004 BC, based on counting the consecutive lifespans of people in the Bible's family trees. We now know from many lines of evidence, discussed earlier in this book, that the Earth has great antiquity, and so does the human pedigree.

If time travel were possible and we could, like the protagonist in H. G. Wells' famous classic *The Time Machine,* move freely through the so-called fourth dimension to land a mere 50,000 years ago in Europe or the United States, we would see a vastly different panorama than exists there today. Huge glaciers would be descended from the north and from the mountains into many of the valleys; the world was then in the grip of the last and grandest of the glacial periods, the Würm in Europe, called the Wisconsinan in North America. Both winters and summers were much colder than at present, and in North America caribou ranged south through Alabama and Georgia, while wooly rhinos were common throughout Europe. The archeological profile of North and South America is barren for this period, but human civilization of a particular and peculiar kind dominated Europe during this time. This is the heyday of the Neanderthal people, named after a valley in Germany near where their remains were first discovered. So-called classic Neanderthals dominated the European continent during most of the last glacial advance. Anatomically modern humans do not appear in the fossil record of Europe until about 30,000 years ago, as the famous Cro-Magnon people, probably responsible for the advanced extraordinary cave art discovered throughout French caves in recent time. But who were the Neanderthals, and of what relation were they to modern humans? The Cro-Magnon culture seems to have replaced that of the Neanderthals very quickly. Does this turnover represent rapid cultural and anatomical evolution, or are these two peoples the evolutionary products of very distinct evolutionary lineages? In this section we examine the fossil evidence that may help us answer these questions and in the process uncover the fossil evidence that links us inexorably to our primate relatives (**Fig. 11.2**).

The title to this chapter includes the term "upstanding" because humans are unique among other Primates in their obligatory bipedal posture, for which many pay dearly with lower back problems later in life. But we are not the only Primates to have ever stood erect, and this is one of the important clues to our heritage in the distant past. We have a long and actually well-documented fossil history, and every year it seems that new fossil discoveries help fill in gaps, refining the minor, and sometimes major, details. For instance, until relatively recently we did not know much about our ancestry in the period back beyond two million years. But this changed dramatically with Donald Johanson's fabulous discoveries of "Lucy" and "The First Family," a treasure trove of hominid fossils more than three million years old from the Hadar region of Ethiopia. Soon after, the late Mary Leakey stunned the scientific world with her discovery in east Africa of 3.5 million year old human footprints, apparently made by the same hominid Johanson had found. Then,

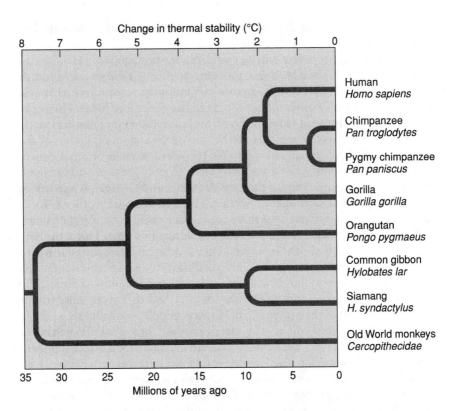

Figure 11.2 The relationships and estimated dates of divergence among living primates result-ing from DNA-DNA hybridization studies. The split between chimpanzees and humans estimated by this molecular method is consistent with current results from the fossil record. (Adapted from C.G. Sibley and J.E. Ahlquist, 1984. The phylogeny of primates as indicated by DNA-DNA hybridiza-tion. *J Mol Evol* 20: 2–15.)

in 2002, Michel Brunet and colleagues announced the discovery of *Sahelanthropus* from Chad, and all our theories on hominid genealogy quickly changed. These are great stories; however, before we delve in detail into our family tree, let us take a quick look and see who is lurking in the branches.

Before Adam: Living and Fossil Monkeys and Apes

Modern Primates
Primates live today in most tropical regions in both the Old and New World. In very general terms, we can identify living prosimians, monkeys, apes, and humans. Prosimians include the lemurs and aye-ayes of Madagascar, the

lorises and bush babies of Africa and southeast Asia, and the tarsiers of southeast Asia. The tree shrews, family Tupaiidae, are sometimes included within the Primates but more often these days are lumped with the Insectivora, a group including all kinds of living and extinct shrew-like placental mammals. Prosimians display some features allying them with other primates, such as an enlarged optic region of the brain and greater reliance on vision, grasping hands and feet, and at least some digits with nails instead of claws, but they lack the completely enclosed bony orbit (eye socket) of monkeys, apes, and hominids (**Fig. 11.3**). A number of them also eat insects as a regular part of their diet, and some are exclusively nocturnal.

The New World monkeys differ from their Old World relatives in having prehensile tails, nostrils that are directed forward (the "platyrrhine" condition), and a dental formula that includes three premolars and three molars. Some New World monkeys include the marmosets, spider monkeys, and capuchins. They are not ancestral to hominids, and we do not consider them further here. Old World monkeys lack a prehensile tail, have nostrils that are directed downward (the "catarrhine" condition), and have only two premolars in their upper and lower jaws. Characteristic Old World monkeys include the macaques and baboons.

Relationships among the apes and humans are becoming better understood thanks in part to the field of molecular biology. Biologists have developed a method in recent years called DNA hybridization, whereby the DNA of one kind of animal can be compared, sequence by sequence, with that of another. It amounts to lining up one half of a chromosome pair of one animal with half from another and seeing if they will bind together. The extent to which they are the same will determine how much of one strand combines with the other. This procedure demonstrates conclusively that human and

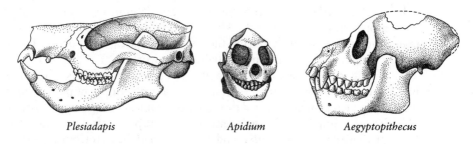

Plesiadapis   *Apidium*   *Aegyptopithecus*

Figure 11.3 General trends in facial contours of early primates and their relatives. Although *Plesiadapis* is not now considered a primate, it expresses a generalized frugivorous/herbivorous dentition and the primitive condition of an orbit not enclosed by bone. *Apidium* and *Aegyptopithecus* are arboreal *Oligocene* primates (anthropoids) from the Fayum Depression of Egypt. Note orbits facing forward and enclosed by bone. (From B.G. Campbell, and J.D. Loy, 1995. *Humankind Emerging*, 8th ed. New York: Allyn and Bacon, p. 180.)

chimpanzee DNA is 98% compatible, whereas the DNA of other primates is less so. There is no explanation for this observation other than a shared genealogy; the chimpanzee and a closely related species, the bonobo, are our closest living relatives. Genetic and paleontological evidence suggests that the chimpanzee-human divergence took place between six and eight million years ago. Further recent fossil discoveries have led to a shift in ape and human classification, so that now the great apes, represented by the orangutans, gorillas, and chimps, and humans are included in the family Hominidae. Because the orangutans split off early from the hominid tree, they are now considered as the sole members of the subfamily Ponginae (the not so much older literature used to include the gorillas and chimps in this subfamily). All the African apes and humans are placed in the subfamily Homininae. A further breakdown, theoretically representing the correct phylogenetic reconstruction, would ally all the bipedal hominines in the tribe Hominini. The colloquial form for this is the awkward "hominin." So, at least in the most recent literature, advanced bipedal hominids are the hominins. The gibbons and siamangs have long been restricted to their own family, the Hylobatidae, and this will not change in the new classification scheme.

Extinct Nonhominin Primates

The earliest primates are difficult to distinguish from other primitive mammals without having complete skulls, which are rare finds in deposits greater than 60 million years ago (Ma). A few paleontologists consider teeth of *Purgatorius,* an early Paleocene genus known only from isolated teeth collected from ant hills in Montana, as those of an early primate, but many other scientists do not find the dental morphology of *Purgatorius* compelling evidence for the appearance of primates at this time. The earliest undisputed primates come from a site in Morocco, Adrar Mgorn, dated around 60 Ma (late Paleocene). *Altiatlasius* was very small, perhaps 75 grams (3 ounces) and is known only from isolated teeth. Although rare during the Paleocene, primates radiated extensively during the following epoch, the Eocene. Forty Eocene genera are known from throughout the world.

Wyoming is not a state one normally associates with primates other than humans in parkas, but about 55 Ma it was a radically different place than it is today. In fact, primates moved through lush tropical or subtropical forests in both north temperate North American and Europe at this time. Similar to living prosimians, ancient *Cantius* and *Teilhardia* had the characteristic stereoscopic vision of living primates and nails instead of claws. *Cantius* was a member of the family Adapidae, whereas *Teilhardia* was an omomyid. Adapids are likely related to the living lemurs, and omomyids seem closer to tarsiers and the monkeys, apes, and hominins.

The so-called higher primates (the anthropoids in some classifications; basically the monkeys and their descendants) diversified during the Oligocene but certainly evolved from prosimian stock during the Eocene. *Eosimias* from China and *Algeropithecus* from Algeria in Africa are examples of Eocene anthropoids. *Eosimias* is dated at about 45 Ma. Like *Altiatlasius, Eosimias* was tiny (maybe 75 grams) and displayed a combination of primitive prosimian skull and dental features along with some characters of the more advanced anthropoids. Other Eocene anthropoids include *Amphipithecus* and *Pondaungia* from Burma and *Siamopithecus* from Thailand. Almost all the relatively small Eocene prosimians from temperate latitudes became extinct after a global cooling event during the early Oligocene known as the Grand Coupure.

Elwyn Simons of Duke University has been collecting primate fossils for many years in the Fayum Depression, a large depositional basin in Egypt, about 100 km southwest of Cairo. Although now one of the most arid regions in the world, during the time primates lived there it was a lushly forested area on the southern border of the Mediterranean Sea. The Fayum beds span the Eocene–Oligocene transition and have produced a number of early anthropoids, including the famous *Aegyptopithecus. Aegyptopithecus* and its relative *Propliopithecus,* another Fayum genus, lie very near the base of the Old World monkeys and the split from monkeys to apes. Some investigators believe that *Aegyptopithecus* may be the earliest ape. One of the ways in which living monkeys and apes are distinguished is by the pattern on the surface of their molars (**Fig. 11.4**). Monkeys have four cusps, whereas apes and early hominins have a five-cusped, or a "Y-5" cusp, pattern. Both *Aegyptopithecus* and *Propliopithecus* have the Y-5 cusp pattern, but in most other regards of the skeleton, these animals were monkey-like rather than ape-like, with a distinctly quadrupedal (four-footed) posture, a relatively long tail, and monkey skull features. Apes also have longer forearms than hindlimbs, a wide chest, short lumbar region of the backbone, and wide pelves. The four-footed arboreal posture is also known as a *pronograde* mode of locomotion, as compared with the *orthograde* mode displayed by apes, in which the spine is held vertical during movement through the trees or on the ground.

The earliest apes are found in African early Miocene sites, about 20 Ma. During the early Miocene, the African plate collided with Europe and western Asia. India collided with southern Asia, and the Himalayan uplift began. It was obviously a very active geological period. With regard to primates, one could call the Miocene the "age of the apes," because 19 genera are known from this period from Africa, Europe, and Asia. I do not list all the genera here, but the ones that concern us most are the African *Proconsul* (early Miocene), European *Dryopithecus* (middle-late Miocene), and African *Kenyapithecus* (middle Miocene). Each of these represents a different anatom-

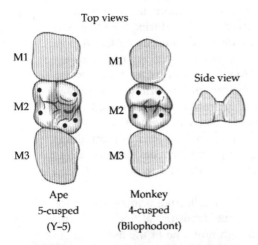

Top views

M1

M2

M3

Ape
5-cusped
(Y–5)

M1

M2

M3

Monkey
4-cusped
(Bilophodont)

Side view

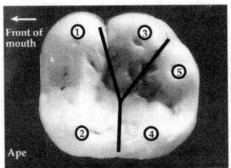

Front of
mouth

Ape

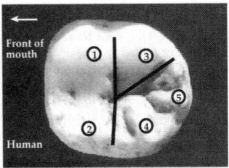

Front of
mouth

Human

Figure 11.4 Ape and human upper molars share a 5-cusped, or "Y-5" pattern, whereas monkeys have a 4-cusped pattern. The two pairs of cusps in monkeys are arranged in anterior and posterior shearing ridges, or lophs (the bilophodont condition). (From B.G. Campbell, and J.D. Loy, 1995. *Humankind Emerging*, 8th ed. New York: Allyn and Bacon, p. 96.)

ical stage in the evolution toward basal hominins. *Proconsul* was an arboreal quadruped but had an ape skull. *Dryopithecus* (from Can Llobateres, Spain) had, in addition to a primitive ape-like skull, the long arms and large hands of a brachiator. Brachiation, or hand over hand swinging through the trees, is believed to have been an ancestral arboreal behavior pattern leading to hominins. This is reflected in a number of anatomical similarities we modern humans share with apes and extinct hominins, particularly the relatively long arms. According to Monte McCrossin of Southern Illinois University, *Kenyapithecus* may be the common ancestor of gorillas, chimps and bonobos, and hominins. *Kenyapithecus*, from east Africa, is dated at about 14

Ma, and according to McCrossin its skeletal anatomy indicates that it was already partly terrestrial.

This leads us into the final approach to modern humans: the appearance of our closest relatives, the extinct African hominins. So what should we see in an ape that is related to modern humans? A short list would include the following:

1. Relatively large size;

2. Upright posture as indicated by ventral rather than posterior position of foramen magnum (hole through which runs the spinal cord) on base of skull, anatomy of leg joints, and anatomy of the lower foot;

3. Relatively large brain and skull;

4. Dental arcade (palate and upper teeth) that is parabolic in shape rather than rectangular, with reduced canines;

5. Relatively flat face;

6. Relatively small teeth with somewhat thickened enamel.

As we shall see, there is not just one species link between apes and modern humans. Evolution did not progress in a straight line from a single ape to a single human species (**Fig. 11.5**). Rather, as the fossil record now amply attests, there were many experiments in being human and many species that combined various ape and human skull and skeletal combinations. Below, the fossil record is reviewed chronologically. The reader will find, as does the author, that there is little agreement among paleoanthropologists on the evolutionary position of a number of species, especially the early ones such as *Sahelanthropus, Ardipithecus, Orrorin,* and *Kenyanthropus.* These are genera making the transition from ape to human, and it may be that most are not on the line to modern humans but are the "experiments" noted above. But the fuzziness of relationships also may simply relate to the way science works. These fossils have been described only within the past few years, and it takes time for detailed anatomical comparisons to be made.

Earliest Hominids

Sahelanthropus from Chad (6.0 to 7.0 Ma)

Until fairly recently, the search for human ancestors was concentrated in eastern and southern Africa. As seen in the sections below, most fossil finds that bear on our ancestry in the twentieth century came from the Great Rift Valley of east Africa and the limestone mines of South Africa. However, for a number of years an international team of paleontologists, led by Michel Brunet of the University of Poitiers, France, has been combing the desolate windblown

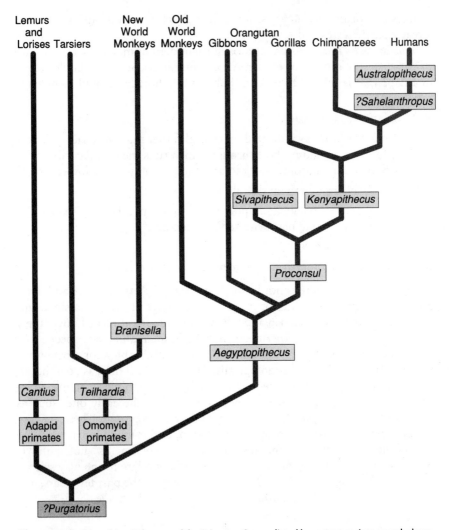

Figure 11.5 A possible phylogeny of the Primates. Genera listed have appropriate morphology to lie in ancestral position but may have been only one among a number of related taxa that flourished at the same time. To the extent that mosaic evolution dominates in organismal evolution, we can expect there to be many experiments in morphology occurring throughout life's history. (Adapted from B.G. Campbell, and J.D. Loy. 1995. *Humankind Emerging,* 8th ed. New York: Allyn and Bacon, p. 192.)

plains of central Chad, 2,500 km from the Rift Valley, in hopes of finding fossil remains that would prove that hominins were not confined to eastern Africa. Brunet had stunned the anthropological community a few years earlier with the discovery of *Australopithecus bahrelghazali* from Chad (see next

Figure 11.6 The skull of *Sahelanthropus tchadensis,* possibly the oldest known hominid, from deposits in Chad that likely range between 6 and 7 million years ago. Although Michel Brunet and his team believe *Sahelanthropus* to be a hominid, others contend that it is an ape and not on the line to modern humans. (Courtesy of the Mission Paleoanthropolgique Franco-Tchadienne.

section), tentatively dated at 3.0 to 3.5 Ma. But the latest find, described as *Sahelanthropus tchadensis* in the July 11, 2002 issue of the science magazine *Nature,* has the scientific community even more excited, because the skull of this hominin expresses a mosaic of features that throws some doubt into our previously held ideas of human ancestry (**Fig. 11.6**).

    Sahelanthropus displays several hominin features, including small canines, relatively thick dental enamel, a relatively flat face, a pronounced brow ridge, and a ventral position of the foramen magnum in combination with a few more typically found in apes, such as small brain size and teeth. The inferred position of the foramen magnum indicates that *Sahelanthropus* likely walked upright (bipedal locomotion). At first glance, the face of *Sahelanthropus* appears to be flatter than that in *Australopithecus afarensis,* a hominin from Ethiopia more than three million years younger that has been considered as a general ancestor for later hominins, including us. So where does *Sahelanthropus* fit in the grand scheme of human evolution? If a hominin, it indicates that the split between apes and humans occurred before about 6.0 Ma. This date is a bit fuzzy because there are no dated ashes in the Sahel region of Chad and the age of the fossils has to be inferred by the evolutionary status of the fossil animals preserved with the hominin (the biostratigraphic approach to dating discussed in Chapter 1). These associated animals further indicate that *Sahelanthropus* lived in or near a lush forested environment. Recently, a group of paleoanthropologists, including Brigette Senut, Martin Pickford, and Wilford Wolpoff, challenged the position of *Sahelanthropus* as a basal hominin, claiming instead that it was female ape.

Kenya, Ethiopia, and Tanzania (6.0 to 3.0 Ma)

In the barren dry wastelands of central Ethiopia and east Africa lie a series of valleys, the edges of which are moving away from each other. This process, driven by the powerful molten forces of the Earth's core, has been splitting eastern Africa from the rest of the continent for millions of years. These depressions are part of the Great Rift Valley that extends inland from the Red Sea, running south through Kenya, Uganda, Rwanda, Burundi, Tanzania, and Mozambique, exiting to the Indian Ocean opposite Madagascar. In the distant future, because two continental plates are moving apart here, the Red Sea will connect with the Indian Ocean, and much of the countries noted above will form their own island continent. It is precisely because this region is so geologically active that we are fortunate to have an almost continuous fossil record of the hominids that lived there and the ability to accurately date their remains, because seismic activity has ripped apart and exposed ancient lake and volcanic sediments containing fossils (**Fig. 11.7**).

Fragmentary remains of a six million year old primate, *Orrorin tugenensis*, were described from Kenya in 2001 by a French and Kenyan team. Found in 2000, *Orrorin* has been informally dubbed "Millenium Man." Brigitte Senut of the National Museum of Natural History in Paris and colleagues consider this Miocene form a bipedal hominid on the direct line to modern humans, but as noted by Michael Lemonick and Andrea Dorfman in their story in the July 23, 2001 issue of *Time,* this conclusion is highly controversial. The evolutionary position of *Orrorin* with respect to *Sahelanthropus* and *Ardipithecus* (see below) remains obscure, and detailed comparisons between these taxa are needed to determine their relationships to each other and to other hominids.

More than five million years ago lush gallery forests existed in the Middle Awash region of north-central Ethiopia, possibly giving way to rich savannas. Volcanos were active along the entire extent of the Great Rift, as indicated by many layers of extruded ash that can be found at various layers in ancient sediments of this region. Because of the argon gas that gets trapped in volcanic glass, these sediments can be dated accurately (see discussion on potassium/argon dating in Chapter 1). Between about 5.8 and 4.4 Ma a hominin by the name of *Ardipithecus ramidus* died, and presumably lived, in

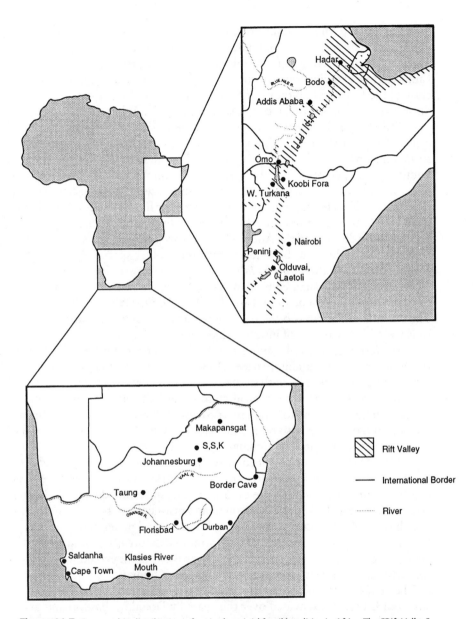

Figure 11.7 Geographic distribution of major hominid fossil localities in Africa. The "Rift Valley" represents a zone on the Earth's surface where two plates are moving apart. S, S, K = Sterkfontein, Swartkrans, and Kromdraai. Klasies River Mouth Cave is one of the earliest sites in the world to produce archaic modern humans, at perhaps 130,000 years B.P. (From I. Tattersall, 1995. *The Fossil Trail: How We Know What We Think We Know about Human Evolution.* New York: Oxford University Press, p. 78.)

the forests of the Middle Awash region. At various times, a large lake existed in the area. The ecological context in itself is interesting, because most early published scenarios of hominid evolution suggested that apes took to the savannas before evolving toward humans. Although the evidence is not yet conclusive, the forested habitat of both *S. tchadensis* and *A. ramidus* certainly throws serious doubt into these early speculations. Tim White, of the University of California at Berkeley, who described *A. ramidus*, labeled it a hominin because of certain skull, dental, and skeletal features that appear to demonstrate kinship with humans rather than with modern chimps and bonobos, but the overall perspective of this creature, if common names were to be applied, would be something more like "the chimpanzee with small canines that stood erect." Two fossil populations of *A. ramidus* have been discovered, one from the Aramis area and another from Addis Ababa. Yohannes Hailie-Selassie, who discovered the Addis Ababa fossils as a graduate student at the University of California, Berkeley, named these older specimens (5.6 to 5.8 Ma) as *Ardipithecus ramidus kadabba*. The somewhat younger material from Aramis (ca. 4.4 Ma), named by Tim White, is considered a slightly more derived member of the same species and is called *Ardipithecus ramidus ramidus*. This "trinomial" form is often used by scientists to indicate that the organisms were likely capable of interbreeding with each other.

If we were able to see *A. ramidus* in the flesh, it probably would not look much different from a chimpanzee. However, a specialist could tell, by its slightly smaller flatter face with smaller canines and its likely habitual bipedal posture, that this was not a typical chimpanzee. The actual position of *A. ramidus* in human prehistory remains obscure. It is certainly near the chimp-human split, but it may also represent a side branch, or "sister group," to the line leading toward modern humans.

In 1995 Meave Leakey described a new species of hominin, *Australopithecus anamensis* from Kanapoi, an area near Lake Turkana in Tanzania. Twenty-one fossil specimens were in the original cache, including dental, skull, and skeletal elements. This species was small and distinct from both *Ardipithecus* and the later *Australopithecues afarensis*. Postcranial material demonstrates that *A. anamensis* was bipedal, and its primitive nature is revealed by its relatively large canines. A reduction in canine size is one of the evolutionary trends characterizing the overall evolution from apes to humans.

The next chapter in our history can be partly described by additional fossils from Ethiopia, excavated over the years by Donald Johanson and associates. Johanson visited the Hadar region along the Awash River in 1973 as part of an international team including Maurice Taieb from France. Their results were sparse but encouraging, and in 1974 Johanson and his international party made an exciting discovery: the almost complete skeleton of a hominid more than three million years old. Inspired by the Beatles song "Lucy in the

Sky with Diamonds" that was playing on cassette as they discussed their find, they named the skeleton "Lucy" (**Fig. 11.8**). Johanson and associates would go on to discover many more skeletons of the same species, *A. afarensis*, named after the Afar people and the region of Ethiopia they occupy. One aggregate of at least 13 individuals apparently died together and has been dubbed "The First Family." These specimens come from localities that have also been bracketed in time by volcanic ash dating, and they range from about 2.8 to 3.3 Ma (**Fig. 11.9**). The discoveries of Lucy and The First Family cleared up a major controversy that had existed in the scientific literature for almost a century. Many individuals claimed that the upright bipedal posture of humans evolved in concert with the enlarged brain and skull that characterizes humans or even that the brain had gotten larger before bipedal status had evolved. Bipedality, they theorized, came late in our evolutionary heritage. Johanson's discoveries completely squelched those ideas. For here was an ancestral hominid more than three million years old, fully erect, approximately the size of a chimpanzee, with the skull and brain size of an ape. It was a startling discovery and a startling revelation.

As is typical among the apes and humans, *A. afarensis* displayed considerable sexual dimorphism; males weighed more and were apparently taller than females. Males may have reached 5 feet in height and weighed up to 50 kg, whereas females would have been closer to 3.5 feet tall and weighed 25 to 30 kg. The foot bones were essentially modern in aspect, though the ends of the toes curved downward more than ours do. (Some scientists have suggested *A. afarensis* was not habitually bipedal, but the foot was so obviously modern in all other aspects that this is very unlikely. Further, as we shall see, there are footprints that provide the "smoking gun" of bipedality.) The great toe of the foot, opposable and used for climbing in chimps and other apes, was instead in the same position as in humans. The hand and wrist bones showed some similarity to those of apes, being slightly curved, but the fingers were not long as in apes and the thumb was fully opposable, as it is in humans but not in apes. The proportions of the upper and lower limbs retained a somewhat primitive condition; relatively short and stout leg bones were combined with relatively long arm bones.

The skull of *A. afarensis* was small, with a cranial capacity of only 380 to 500 cc (humans average about 1350 cc), and the facial region closely resembled that of a chimpanzee (**Fig. 11.10a**). The dentition was intermediate between apes and humans in a number of ways. Of particular significance is the form of the canines that in chimpanzees are large and pointed. The lower canine in chimps is so large that a space, called a diastema, is present in the upper jaw to contain it. In *A. afarensis* the canines are much smaller but still somewhat pointed and maintain a stout root; a small diastema is present in the upper jaw. In humans the canine is smaller still, and the diastema has

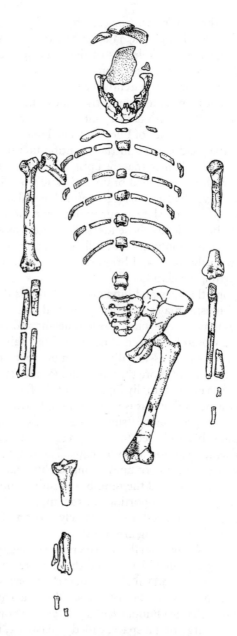

Figure 11.8   The skeleton of *Australopithecus afarensis* ("Lucy") from Hadar, Ethiopia.   Donald Johanson's famous discovery showed conclusively that an early hominid with a small brain walked upright.] (From I. Tattersall, 1995. *The Fossil Trail: How We Know What We Think We Know about Human Evolution.* New York: Oxford University Press, p. 143.)

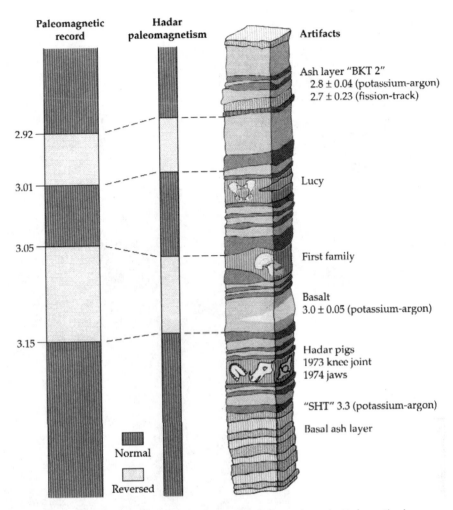

Figure 11.9 Stratigraphic relationship of important fossil discoveries at the Hadar section in Ethiopia that produced numerous remains of *Australopithecus afarensis*. Note that the sediments containing fossils are bracketed by a number of radiometric dates. (From B.G. Campbell, and J.D. Loy. 1995. *Humankind Emerging*, 8th ed. New York: Allyn and Bacon, p. 250.)

been lost. The first lower premolar is also exactly intermediate in form between apes and humans. Jaw anatomy is also illuminating. In chimpanzees the front of the jaw is thick and the cheek teeth (premolars and molars) are directed backward in two parallel lines behind the large canines. Together, the lower teeth are said to form an "arcade." This arcade has a different form in *A. afarensis* and in modern humans. In humans the arcade is more rounded and is wider at the back (toward the molars) than at the front (toward the

(a)

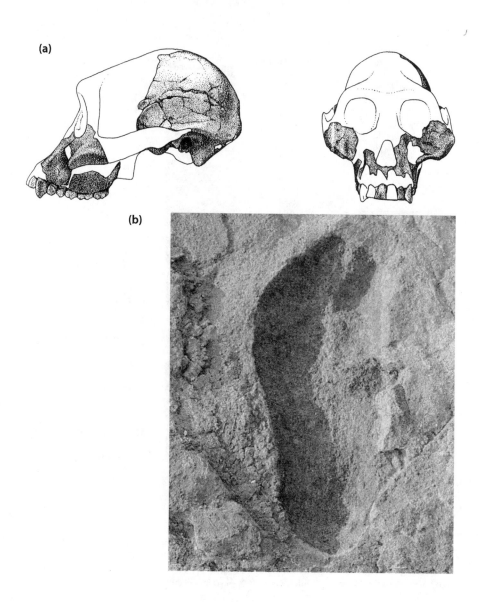

(b)

Figure 11.10 *Australopithecus afarensis.* (a) Reconstructed skull from Hadar, Ethiopia, (b) foot-print preserved in fossilized volcanic ash from Laetoli, Tanzania. The hominid footprint is about 3.6 million years old. ([a] from I. Tattersall, 1995. *The Fossil Trail: How We Know What We Think We Know about Human Evolution.* New York: Oxford University Press, p. 146; [b] from B.G. Campbell, and J.D. Loy, 2000. *Humankind Emerging,* 8th ed. New York: Allyn and Bacon, p. 258.)

canines). The jaw is also much thinner at the symphysis in humans (where both halves fuse during development; the chin region). *Australopithecus afarensis* is almost perfectly intermediate in these features.

The holotype (or just "type") specimen of *A. afarensis* actually comes from another country, Tanzania. When they named the new species *A. afarensis*, Donald Johanson and Tim White chose an exemplary lower jaw collected by Mary Leakey, from an area named Laetoli, as the type specimen. This turned out to be providential, as Paul Abell, a geochemist working at Laetoli with Leakey in 1978, discovered footprints made by this hominid in volcanic ash dated to about 3.6 Ma (**Fig.11.10b**). In fact, two individuals are represented, and the trail goes for over 24.4m (80 feet). During this brief walk the two individuals, one slightly larger than the other, stopped and turned at one point, as if to look at something off in the distance, and then continued on their way. The imprints are entirely human in shape, with a deep "strike" for the heel and a raised arch. The big toe is clearly in line with the other toes. Remarkably, many other large animals also passed through the same area of volcanic mud about the same time; additional tracks were made by elephants, rhinos, antelopes, giraffe, and a saber-toothed cat.

A recent australopithecine discovery in Chad, north-central Africa, has paleoanthropologists excited because the locality, Bahr el Gazal, is so far from the Hadar region of Ethiopia and the Great Rift Valley, attesting to a widespread distribution of early hominids. The material has been named a new species, *Australopithecus bahrelghazali*, by Michel Brunet and colleagues. Although no radiometric dates are available for this hominin, on biostratigraphic grounds Brunet places *A. bahrelghazali* between about 3.5 and 3.0 Ma.

The early record of hominins took another unexpected and interesting turn when Meave Leakey and colleagues described yet another new form from east Africa in 2001. *Kenyanthropus platyops* apparently lived west of Lake Turkana, Kenya, between about 3.3 and 3.5 Ma. Like other early hominins, it was clearly bipedal and demonstrated a curious combination of primitive and derived skull features. The defining feature for this hominin was its unusual flattened face (thus "*platyops*" in the species designation), combined with relatively small molars (**Fig. 11.11**). Its phylogenetic position in the hominin family tree is not certain at this time, but Meave Leakey has noted certain resemblances to the later *Homo rudolfensis*.

Subsequent to three million years there was apparently a considerable radiation of hominins, and the nomenclature of all the forms is controversial. Some scientists, who tend to be "splitters," believe there might have been many different species of hominins living simultaneously in Africa and elsewhere throughout the late Pliocene and Pleistocene. Others, particularly those who do not recognize phyletic speciation, take a more conservative approach

Figure 11.11 Skull of *Kenyanthropus platyops*, an enigmatic hominid that lived in east Africa, near Lake Turkana, about 3.3 to 3.5 Ma. (Copyright National Museums of Kenya.)

and suggest that hominins were capable of wide geographical distribution in very brief periods of geological time and that gene flow was maintained among those populations whose anatomy is basically similar. These "lumpers" recognize fewer endemic species of hominins. I will endeavor to reflect these different positions throughout this chapter. Because there is considerable overlap between the species in time, in the sections below I consider each of the post-3.0 million year anatomical and ecological types separately, especially if it is known primarily from one geographical region.

Australopithecus africanus (2.8 to 2.5 Ma)

Far to the south of Ethiopia, deep in the land of the Zulu and the Boers, is a dissected limestone plain mined by modern people since the turn of the century. Lime is a prime fertilizer, and in South Africa it was also used in gold refining. Worldwide, limestone mines continue to produce treasure troves of fossils, if there is someone either in the mining industry or in paleontology that is savvy enough to recognize fossils when they turn up and is prepared to excavate on a moment's notice. When I was a graduate student at the University of Florida in the late 1960s, I spent a good deal of my time walking limestone mines looking for the tell-tale signs of bone weathering out of reddish-brown cave fillings that were exposed by the miners. Amateur fossil hunters in the Gainesville area often helped, and through these efforts new fossil localities are added each year. But one must be diligent. The mining engineers will not hold up their digging for long. Many times the University of Florida staff has had to take only a sample from a fossiliferous locality before the complete site was mined out. It is the sad truth that we probably lose thousands of critical fossils each year through mining operations, but we also gain many that never would have been exposed otherwise. In a sense, this was also the brief story of Raymond Dart, Professor of Anatomy at the University of Witwatersrand, and the Taung child, the world's first real "missing link" in hominid evolution.

According to Ian Tattersall in his book, *The Human Odyssey*, Dart was handed a box of fossils as he was dressing to be best man at a friend's wedding. Although fossils were known from the Taung limeworks, the box contained the facial region and natural braincast of a hominid child. Dart described the specimen in 1925 as *Australopithecus africanus*, "African southern ape," and championed it as an intermediate between apes and humans. However, the scientific community did not respond to this announcement positively, primarily because Dart's discovery, suggesting a small ape-like creature that walked erect, did not fit with the model they had been led to accept through the discovery of another so-called missing link, Piltdown Man. This find was represented by a complete skull unearthed in the Piltdown section of Sussex in 1912. Subsequently determined to be an elaborate hoax, until the fraud was exposed in 1953 the features of the Piltdown hominid were accepted as those likely in our ancient ancestor. The skull of Piltdown was large, essentially modern in anatomy, but the lower jaw and presumably part of the reconstructed face was ape-like. This combination of features suggested the brain became enlarged early in our heritage, and consequently the European, and particularly the British, scientific community was not ready to accept a place for Dart's diminutive and decidedly ape-like fossil in the human record. Unfortunately, Dart was not inclined to determine where the Taung fossils originated, and subsequently the locality was destroyed. However, adult specimens of *A. africanus* turned up at nearby sites known as Sterkfontein and Makapansgat in 1936 and 1947, respectively, and proved conclusively that Dart was correct in his assessment. Although dating the South African apemen is difficult because no volcanic ashes are available in the Transvaal region, a number of independent lines of evidence point to a possible range in time from about 2.5 to 2.8 million years for *A. africanus*.

The field and laboratory work of a number of intrepid South African fossil hunters, most notably Robert Broom, John Robinson, and C. K. Brain, show how meticulous collecting and attention to detail can sometimes unravel even the most contentious problems. In addition to the remains of fossil hominids, the South African cave sites also included numerous broken bones of large mammals, both herbivorous and carnivorous (e.g., leopards). Dart interpreted these bones as the refuse of *A. africanus* and also concluded that these little ape-men used the bones and jaws of dead mammals as weapons. There is a famous film clip of Dart, gesticulating frenetically, acting out the brutal way in which these early hominids likely battered their own kind as well as other animals. Subsequent research has shown that rather than predator, early australopithecines appear to have been the prey of leopards, and the bone collection seems to have been the handiwork of porcupines. Many rodents accumulate and gnaw bones, apparently for both the calcium and fat content.

Ethiopia continues to produce fossil hominins, and as with most hominin discoveries, the most recent find from this region of Africa was described as a new species, *Australopithecus gahri*, by B. Asfaw and colleagues in 1997. A series of fossils from about 2.7 Ma, representing parts of the skull and skeleton, point to another hominin that, according to the authors, may be on the route to becoming human. That is, instead of *A. africanus* being ancestral to later hominins, this distinction might fall instead to *A. gahri*. With this evolutionary scenario, the south African *A. africanus*, rather than being a common ancestor for later hominins, might be relegated to a side branch.

The Robust Australopithecines (2.7 to 1.3 Ma)

Depending on the phylogeny of hominids in vogue, the robust australopithecines are classified either with the gracile forms in the genus *Australopithecus* or separately in their own genus, *Paranthropus* (**Fig.11.12**). Currently, the robust species are considered as a separate genus, distinct from *Australopithecus* in presumed ecological choice and morphology. Below, where I may indicate that a particular robust species was named by certain investigators, for consistency I use the generic name *Paranthropus*, even if the authors originally named the species as an *Australopithecus*.

Ethiopia and Kenya (2.7 to 2.3 Ma): *Paranthropus aethiopicus*
In southern Ethiopia, also in the Great Rift Valley, the Omo River drains southward into huge Lake Turkana just over the border in northern Kenya. The deposits in this area were prospected by French paleontologists Camille

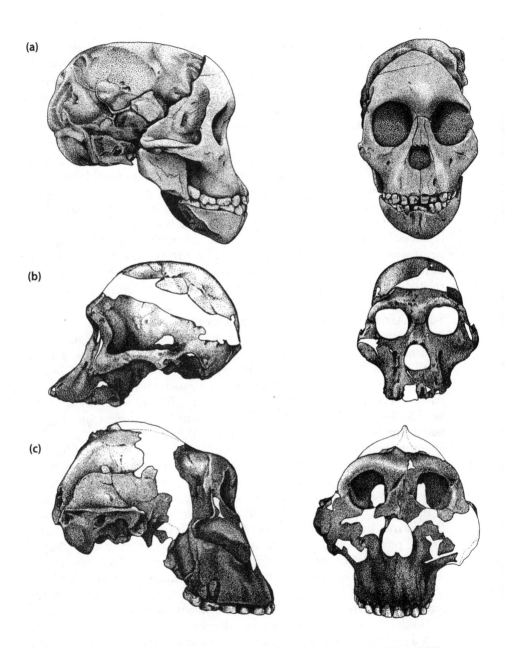

Figure 11.12 Skulls of *Australopithecus africanus* and *Paranthropus boisei.* (a) The "Taung child," South Africa, originally described by Raymond Dart in 1924; (b) an adult *A. africanus* from Sterkfontein Member 4, South Africa; (c) *P. boisei* from Bed I, Olduvai Gorge, Tanzania. (From I. Tattersall, 1995. *The Fossil Trail: How We Know What We Think We Know about Human Evolution.* New York: Oxford University Press, pp. 70, 107, and 56.)

Arambourg and Yves Coppens and the American F. Clark Howell back in 1967, and the area has produced a wealth of fossil mammal material and some hominin specimens. In addition to a species similar to *A. afarensis* at about 3.0 Ma, the research team discovered a "hyper-robust" australopithecine that was clearly not the same species as *A. africanus* or *A. afarensis* and that extends in the Omo outcrops from about 2.5 to 1.2 Ma. A specimen known as the "black skull" recovered from slightly older (2.6 Ma) sediments to the west of Lake Turkana in Kenya is also likely a member of this early form. Because more material of this general kind of hominid is known from the South African sequence, it is described in detail below. The important point here is that this robust type of hominid is found as early as 2.6 Ma, although in Ethiopia it apparently did not overlap in either geographical distribution or time with the gracile *A. afarensis*. It was named by the French team as *Paranthropus aethiopicus*, but it differs only in slight details from the later occurring and obviously closely related forms *P. robustus* and *P. boisei*.

South Africa (2.0 to 1.0 Ma): *Paranthropus robustus*

Robert Broom (Transvaal Museum) is credited with first recognizing the presence of a robust hominid from the Transvaal caves. In addition to the small gracile forms from Taung, Makapansgat, and Sterkfontain, Broom and his associates discovered remains of a much larger hominid, with a flatter face and flared cheekbones, from cave fillings at Kromdraai and Swartkrans. Broom named this hominid *Paranthropus robustus*. The robust form averaged perhaps 68 kg (150 lbs), whereas large adults of *A. africanus* probably topped out at 46 kg (100 lbs). The jaws and teeth of *P. robustus* were huge, and, at least in males, a large bony crest was developed on the top of the skull to anchor the powerful jaw musculature. Crests such as these are common in gorillas, but not in other hominids. Scanning electron microscope imaging of wear patterns on the molars of *P. robustus* by Fred Grine at the State University of New York (Stony Brook) points to an almost exclusively herbivorous diet, whereas that of *A. africanus* was probably more diverse and composed of less fibrous materials. The grinding surface of *A. africanus* molars are worn smooth, whereas those of *P. robustus* are pitted and gouged. The robust australopithecines apparently did not overlap in time with the earlier gracile type in South Africa, as they did not in Ethiopia. Brain and others have suggested that the deposits at Kromdraii and Swartkrans were deposited somewhere between 1.6 and 2.0 Ma. Detailed anatomical studies of Swartkrans *P. robustus* postcranial material by Randall Sussman of SUNY Stony Brook document the likelihood that these robust hominids had a fully upright posture and had the anatomical potential to both make and use stone tools. The thumb was opposable as in modern humans.

Kenya and Tanzania (2.0 to 1.2 Ma): *Paranthropus boisei*

Richard Leakey, then at the Nairobi National Museum in Kenya, began his field season in 1967 with the same team that discovered fossils in the Omo Valley, but he opted instead to leave the group and reconnoiter on his own across the border into Kenya, along the shores of Lake Turkana (earlier known as Lake Rudolf). There, in an area called Koobi Fora, he and his crew discovered numerous remains of robust australopithecines, including almost all of the skeleton. These specimens were referred to *Paranthropus boisei,* named earlier from Olduvai Gorge in Tanzania (originally as *Zinjanthropus boisei*) by Richard's father, Louis. The robust hominid at Koobi Fora apparently ranged in time from about 2.3 to 1.2 Ma.

Louis Leakey was an eccentric energetic paleontologist who worked in the Olduvai Gorge region of Tanzania for most of his adult life. He died in 1972. His wife Mary continued this legacy until her death in 1996. Olduvai Gorge is a great rent in the Earth's crust, a small version of the Grand Canyon. Part of the fossiliferous Great Rift Valley, it has almost a continuous sequence of sediments spanning the period from about 2.2 Ma to modern time. Louis Leakey named *Zinjanthropus boisei*, which he called "nutcracker man" based on its large molar teeth, from a site in the Gorge potassium/argon dated at 1.75 Ma. Robert Broom had earlier suggested a date of 2.0 million year for *Paranthropus robustus* from South Africa, but this date had not been accepted by the scientific community. It was only after an early date was confirmed for the Olduvai robust hominid that Broom's idea gained credibility. *Paranthropus boisei* apparently occupied the shores of a large lake in Olduvai Gorge from about 2.3 to 1.3 Ma (**Fig. 11.13**).

The Earliest Humans (2.4 to 1.60 Ma): *Homo habilis* and *Homo rudolfensis*

During the 1960s Louis and Mary Leakey made another discovery from Olduvai Gorge that was immediately controversial; they claimed to have uncovered the remains of a member of our genus, giving it the name *Homo habilis,* or "handy man," because the fossil material was clearly associated with fabricated stone tools. The Leakeys' detractors argued that the specimens were fragmentary and little different from *A. africanus,* but more than 30 years later the Leakeys have been vindicated, because additional material spanning the time period from about 1.9 to 1.60 Ma in northern Kenya and South Africa confirms the presence of an advanced hominid that made and used tools.

The skull and skeletal anatomy of *H. habilis is* not much advanced over that of *A. africanus* and in fact preserves some of the primitive aspects of the early australopithecines, including the disproportionately long arms relative

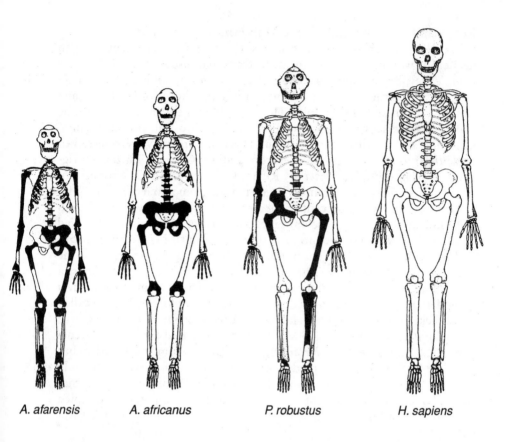

A. afarensis   A. africanus   P. robustus   H. sapiens

Figure 11.13 The skeletons of three australopithecines and a modern human. Dark areas represent the elements preserved as fossils in single skeletons; the remainder is inferred. Skulls are known for all. Skeletal structure and the position of the foramen magnum (the hole at the base of the skull where the spinal cord enters) indicate that australopithecines stood erect. (From *Lucy: The Beginnings of Humankind* by Donald Johanson and Maitland Edey. New York: Simon & Schuster, Inc., 1981. Copyright 1981 by Donald C. Johanson and Maitland A. Edey.)

to the legs. The skull, and therefore by inference the brain, was larger, about 650 cc (*A. africanus* was about 450 cc), and with the associated stone tools, is the dominant factor in recognizing that a species has passed the threshold to our genus. The tools were not very advanced; the "Oldowan" tool culture, as it is called, consists of a central core of stone, often of volcanic origin, off of which a few flakes were broken by a hammerstone. There is continued debate as to whether it was the core or the flakes the hominids were using as tools, though the pendulum seems to have swung toward the flakes, because they show some signs of being further worked, or "retouched," as if to get a finer edge. As noted below, a few specialists believe that *H. habilis* may represent a small species with disproportionately long arms and short legs that was not on the direct line to modern humans.

For a number of years paleontologists have known of significant variation in some hominid skulls found by the Leakeys in east Africa. Dominant among these was skull KNM-ER 1470 from the Koobi Fora region, which had sometimes been considered as a variant of *H. habilis* (**Fig. 11.14**). The British paleoanthropologist Bernard Wood suggested that this skull and some other material from east Africa and Ethiopia should represent a new species, *Homo rudolfensis*, that lived along side *H. habilis* from perhaps 2.4 to 1.6 Ma. This interesting hypothesis, shared by Ian Tattersall of the American Museum of Natural History in New York, suggests that our ancient family tree might have been bushier than originally suspected.

Advanced Fossil Hominins and the Origin of Modern Humans

Sometime after about 1.8 Ma relatively advanced hominids appear in east Africa. These hominids were basically modern in stature and size, and from the neck down one would have a difficult time distinguishing their bodies from our own. In fact, most of the known specimens, usually referred now to the species *Homo ergaster,* fall out into the upper limits of body size even for modern humans. The phylogenetic connection to earlier hominids is not securely established, but Ian Tattersall has suggested that later *Homo* may be descended from *H. rudolfensis* rather than from the diminutive *H. habilis*. The best preserved skull of *H. rudolfensis,* KNM-ER 1470, has a cranial capacity of 750 cc, distinctly larger than that of *H. habilis* (KNM-ER 813; 510 cc).

Lean and Large: *Homo ergaster* and *Homo erectus* (1.9 to 0.5 Ma)

Advanced hominids discussed in this section are known predominantly from southeast Asia and Africa. Originally, all this material was referred to *Homo erectus,* but the African material is now considered to represent a distinct species, *H. ergaster*. Although obviously related, there are differences between the species that indicate the Asian hominin may represent an evolutionary dead end rather than a connection to modern humans as has been proposed in the past, when the African material was unknown. Let us begin with the African material, because it may predate the Indonesian *H. erectus* by more than a million years.

The link between earlier *Homo* and later hominids, including ourselves, is known. Specific fossil evidence are skulls found by Richard Leakey from East Turkana, Kenya dated at about 1.7 Ma. These specimens, including the famous "KNM-ER 3733" cranium, combine some primitive features of *H. habilis* and *H. rudolfensis* with an enlarged braincase (about 850 cc) and in-

(a)

(b)

(c)

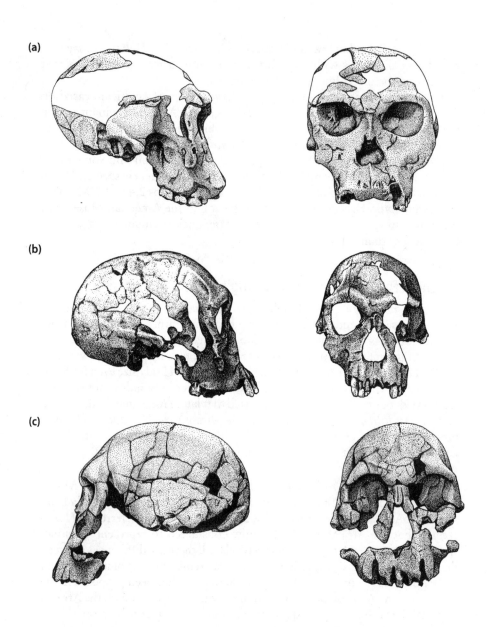

Figure 11.14 Skulls of *Homo habilis* and *H. rudolfensis.* (a) *H. habilis;* OH 24 from Bed I at Olduvai Gorge, Tanzania. (b) *H. habilis;* KNM-ER 1813 from Koobi Fora, Kenya. (c) *H. rudolfensis;* KNM-ER 1470 from Koobi Fora, Kenya. (From I. Tattersall, 1995. *The Fossil Trail: How We Know What We Think We Know about Human Evolution.* New York: Oxford University Press, p. 114, 133, 134.)

cipient features of later *Homo*. Another famous discovery of this intermediate hominin, the "Turkana Boy" (**Fig. 11.15**), about 1.6 million years in age, was recovered in 1984 from Nariokotome, west of Lake Turkana, also in Kenya. Although only about 12 years old, the almost complete skeleton clearly shows that this individual would have likely reached 6 feet in height at maturity. At his death, the boy stood at 5 foot 3 inches and weighed slightly over 100 pounds. He was tall and slender, similar in build to current inhabitants of the area. Because the braincase of these specimens is relatively lightly built as compared with that of later *H. erectus* from Indonesia, Ian Tattersall suggests that both modern humans and classic *H. erectus* might have evolved from these populations as separate lineages. This is, in part, his rationale for naming these early African populations *H. ergaster*. Modern human skulls have relatively thin bone, and Tattersall and colleague Niles Eldredge argued in 1975 that it was more likely modern human skulls would have evolved from late Pliocene thin-walled ancestors than in an evolutionary reversal from thick-walled *H. erectus*. However, other options exist, which are considered below.

C. K. Brain (Tranvaal Museum) and Elizabeth Vrba (now at Yale University) also found remains of advanced hominids at Swartkrans in South Africa in deposits likely 1.6 to 1.8 million years old and possibly contemporaneous with *P. robustus*. This is actually quite curious, because it does not intuitively seem likely the two species would have peacefully cohabited the cave system at the same time. In fact, in the levels where most of the robust australopithecines have been found at Swartkrans, few bones of the advanced hominid are present. Brain has suggested that at least some of the robust australopithecines from Swartkrans were killed by leopards, because puncture wounds in the skull of one specimen match up perfectly with the upper canines of an adult leopard (**Fig. 11.16**). Leopards do not normally live in caves, but they do lounge around with their kills in trees that grow near cave entrances, and it is not inconceivable that stockpiles of bones under the trees would have made their way into the cave, perhaps partly through the hoarding behavior of the porcupines mentioned previously. So it is likely that *Paranthropus* and early *Homo* did not live in the cave simultaneously. Both may have been prey for the leopards. Because the remains of *Homo* become common in the upper levels of the cave about 1.0 Ma, it is conceivable it lived in the cave at this time. Brain has also suggested that this early *Homo* used fire, but this remains controversial. Although the evidence is minimal, the advanced hominin may have overlapped with *P. robustus* briefly before replacing it in this region of South Africa.

Homo ergaster specimens are also found throughout the fossiliferous sedimentary sequence at Olduvai Gorge, Tanzania, from about 1.7 Ma to slightly more than 0.6 Ma. Elsewhere in Africa, they have been recorded in Morocco,

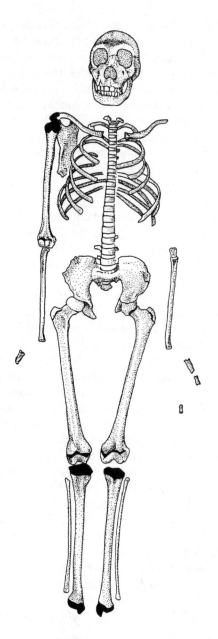

Figure 11.15 The almost complete skeleton of the "Turkana boy," a young (probably male) *Homo ergaster* from Nariokotome, Kenya; about 1.6 Ma. This individual combines a primitive skull with an essentially modern skeleton. He would have stood over six feet tall as an adult. (From I. Tattersall, 1995. *The Fossil Trail: How We Know What We Think We Know about Human Evolution.* New York: Oxford University Press, p. 189.)

Figure 11.16 This disturbing illustration probably reflects the dangerous life of early hominids in Africa. At least one skull of *Paranthropus robustus* has been found at the Swartkrans limestone mine in South Africa with holes perfectly matching the shape and bite width of a leopard. (From I. Tattersall, 1995. *The Fossil Trail: How We Know What We Think We Know about Human Evolution.* New York: Oxford University Press, p. 201.)

Algeria, and Ethiopia, spanning the time from 1.0 to slightly less than 0.5 Ma. But when did these early hominids leave Africa, if they originated there? Dates associated with some new material and new dates from older localities appear to be pushing the time that humans left Africa back considerably. Scanty hominid materials from Longgupo Cave in China, stream deposits in central Java, and ancient lake beds near Granada, southern Spain, are all supposedly older than 1.8 Ma. The most exciting new specimens that bear on the timing of dispersal from Africa come from Dmanisi in the Republic of Georgia. Excavated from beneath a medieval castle by a team led by Leo Gabunia and David Lord Kipanidze, a well-preserved skull and associated fossil material dated to about 1.8 Ma may represent an archaic form of *H. ergaster.* Stone tools found with the skull are of the more ancient Oldowan (pebble tool) type rather than the Acheulean tools associated with *H. erectus* from China and Java.

In the 1890s a Dutch physician, Eugene Dubois, specifically went to eastern Asia to seek out fossil humans. He volunteered for service with the Dutch East Indian Army and while serving in Java discovered the first hominid missing links outside of Africa. Originally named *Pithecanthropus erectus,* the fossil hominins from Java and material of the same species described later from Zhoukoudian in China by the Canadian anatomist Davidson Black are now referred to as *H. erectus.* We do not know for certain when *H. erectus* reached China and Indonesia; it could have been as early as 1.7 Ma. But these early dates have not been substantiated. *Homo erectus* from this region has a skull that exaggerates incipient features seen in the earlier African *H. ergaster* populations. The forehead is low and ends at a pronounced ridge of bone, the supraorbital torus. Another ridge of bone, or torus, on the occipital bone at the back of the skull gives the cranium a characteristic ovoid shape. The frontal bone was low and receded at a low angle. The top of the skull, rather than being round as in modern humans, had a pronounced ridge, though not a distinct crest as in the robust australopithecines. Cranial capacity of the earliest forms referable to *H. erectus* was about 990 cc, ranging to 1100 cc in the latest specimens. The face was relatively short and broad and jutted forward ("prognathism"). The mandible was robust and lacked a distinct chin. The cheek teeth were relatively small compared with the size of the skull, a feature that represents part of a trend ending with even smaller teeth in *Homo sapiens*. The postcranial skeleton was essentially modern in all regards. These features also characterize *H. ergaster* but not to the same extent. Everything in *H. erectus* is exaggerated, more robust, as compared with *H. ergaster*. In particular, the skull bones are much thicker in *H. erectus*. As noted above, these anatomical differences suggested to Ian Tattersall and others that *H. erectus* did not give rise to modern humans. We return to this idea later, because it is not shared by all investigators.

Archaic Humans of Africa, Europe, and the Middle East (post-0.5 Ma)

The hominin skull from Dmanisi provides compelling evidence that humans first exited Africa more than 1.8 Ma. These first explorers, at the *H. ergaster* evolutionary level, then made it as far as Indonesia and either gave rise to a distinct species, *H. erectus,* or evolved relatively gradually into *H. erectus*. But did one or both of these species give rise to modern humans?

A number of advanced hominin skulls have been recovered from all over Africa, the Middle East, and western Europe that bridge the gap between *H. ergaster/H. erectus* and modern *H. sapiens*. They are a diverse lot but have some primitive features in common, including large brow ridges, low frontal bones, relatively thick cranial bones, and relatively robust, chinless mandibles. Advanced features include large endocranial volume (>1000 cc), rounded occipital bones, expanded parietal bones, and more arched brow ridges. Some localities with humans demonstrating this combination of characteristics include Swanscombe (Great Britain), Arago (France), Mauer and Steinheim (Germany), Petralona (Greece), Broken Hill (South Africa), Bodo (Ethiopia), Ngandong (Java), and Atapuerca (Spain). Because many of the fossils are poorly sequenced in time, some of the differences between them probably represent sampling from different time intervals. Among these fossils are possibly representatives of three lineages, one leading to *H. erectus,* another to the classic Neanderthals of late Pleistocene Europe, and a third leading to modern humans. In the recent past, all these specimens were lumped in *Homo heidelbergensis*. Exciting new material from the Gran Dolina cave system of northern Spain, in the Atapuerca hills, may provide a way to interpret some of this material. The deepest layers are dated to about 0.78 Ma and include hominin skull material that is different from both *H. ergaster* and *H. erectus*. In particular, the lower jaws are smaller, the cranial capacity is large (>1000 cc), the brow ridges are separate, and the mid-facial region is, at least in one juvenile individual, relatively flat (not prognathic). The Spanish team that discovered the material has named it as a distinct species, *H. antecessor*, and they suggest it was ancestral to both modern humans and the later Neanderthals. They also suggest that much of the European hominin material referred to *H. heidelbergensis* may represent populations that were on the line leading to Neanderthals, considered as a separate species, *Homo neanderthalensis*. However, we will consider here *H. antecessor* as an archaic form of *H. heidelbergesis*.

Let us summarize this complicated evidence. Putting all the work by various investigators together, we derive an evolutionary scenario for advanced hominins that can be summarized as follows:

1. *Homo* first appears in Africa, as *H. habilis* and *H. rudolfensis*.

2. *Homo ergaster* evolves in Africa, possibly from *H. rudolfensis*.

3. *Homo ergaster* disperses out of Africa about 1.7 Ma.

4. *Homo ergaster* gives rise to at least two lineages, one leading to Asian *H. erectus* and another leading, in southern Europe, to *H. heidelbergensis*.

5. *Homo heidelbergensis* gives rise to two lineages, one leading to the robust *H. neanderthalensis* (= *H. antecessor*) and another leading to *H. sapiens*.

We do not have enough fossil material to say for certain when *H. heidelbengensis* first appeared. We only know that an archaic form was in northern Spain about 780,000 years ago. It is probably fair to say that humans with modern skull characteristics appeared after that time, and now let us examine the fossil material leading both to us and the famous "cavemen" of the European late Pleistocene.

Earliest Modern Humans and Their Neanderthal Contemporaries (0.12 to 0.03 Ma)

It is probably not accidental that modern humans first appeared during the Pleistocene. The Pleistocene, or Ice Ages, was certainly not the first time the world was ravaged by great glaciations, because there is evidence for glaciation back in the Paleozoic Era. In both cases we believe that the history of life was changed because of the pronounced effects these colder periods had on the biosphere. The Pleistocene changes would have been most obvious at the higher and middle latitudes. For instance, at the height of the last major glacial advance, called the Wisconsinan in North America, most of Canada and Alaska were covered by an ice sheet and glaciers made their way down the Rockies, filling the valleys that give them their craggy and sharp-bluffed appearance. The same was true in the Alps of Europe. These mountain glaciers are called alpine glaciers, in contrast to the larger continental glaciers that represent the expansion of the entire Arctic ice cap. The glaciers advanced and retreated numerous times in the past 1.8 million years, at least nine times in the last 800,000 years alone (it may not be coincidence either that the beginning of the Pleistocene coincides approximately with the first dispersal of humans outside of Africa). The glacial advances were coupled with sea level retreat, sometimes of a staggering magnitude. If you were able to stand on top of the Empire State Building in New York 20,000 years ago, looking to the east as far as you could see would be a large grassy plain with rivers running to the sea and probably trees in copses along the water courses. From this vantage point, with a pair of binoculars, you could probably see herds of mammoths and bison, native horses, and perhaps a small herd of musk oxen, all foraging where now the Queen Mary comes into its berth. The giant beaver *Castoroides*, about the size of a small black bear, might have been frolicking in the water. If you were lucky, you might also see either the native American lion or a saber-toothed cat sneaking up on its prey. It is difficult to say if you would have been able to see the ocean—perhaps from the top of the Empire State Building—but certainly not from the ground. In Europe, it would have been a tough time for humans to survive, and at the very least they would have needed clothing and shelter and fire.

Neanderthals: Europe's Cave People

Prehuman fossils are not new to science, but their acceptance was not instantaneous. The earliest records of fossil hominins are likely those of various Neanderthals from caves in Belgium and the Neander Valley (Thal) in Germany. All were found before 1860, but it was not until 1864, when William King, a professor of Anatomy at Queen's College in Ireland, named *H. neanderthalensis* for another skull from a cave in Gibraltar, that the European scientific community began to seriously consider the possibility that ancient humans had inhabited the planet. However, following King's report, a German pathologist named Rudolf Virchow convinced King and other scientists that the skull from the Neander Valley had come from a Prussian soldier, old and suffering from arthritis and an earlier childhood bout with rickets, who had stumbled into the cave and died. This was in part a response by creationists of the time, still unwilling to accept the evidence for evolution presented by Charles Darwin in his recently published book, *The Origin of Species* (1859). However, by the end of the nineteenth century there were just too many hominin fossils to ignore. Neanderthals, in association with extinct animals and stone tools, were cropping up everywhere in Europe, and Eugene Dubois had found *H. erectus* in Asia.

So-called classic Neanderthals were large-boned powerful people, though on average they were not as tall as modern humans (**Fig. 11.17**). Their brains were large, but their skulls still retained some of the primitive features inherited from *H. ergaster* or *H. heidelbergensis,* such as a somewhat ovoid skull and massive (though individualized) brow ridges. Their forehead was also relatively flat as compared with ours. As in *H. ergaster* and *H. erectus,* the lower jaw lacked a well-defined chin. A distinct space, called the retromolar gap, is found behind the last molars. The face was somewhat prognathic, and the nostril openings were much larger than ours. In the mid-twentieth century, Neanderthals were considered to represent an intermediate in a link from *H. erectus* to *H. sapiens,* but with many recent fossil hominin discoveries, that possibility is less and less sponsored by professionals. These days, both *H. erectus* and *H. neanderthalensis* are considered as separate specialized species that became extinct.

The geographical range of Neanderthal fossils in Europe shows they were confined to rocky areas with caves, and there are no certain finds of Neanderthals outside of caves (**Fig. 11.18**). This is almost certainly why they did not make the trek through Siberia and into North America across the Bering land bridge, as did many other mammals during the Pleistocene when sea level was low. Both the oldest and the youngest Neanderthal remains may come from Spain. The same cave system at Atapuerca that produced *H. antecessor* has also produced, from Sima de los Huesos (Pit of Bones), hominin skulls that include a melange of features, including some that show unmistakable resemblance to Neanderthals. Another cave site near the town of Zafarraya in southern Spain has produced stone tools and Neanderthal fossils that date in the range of 27,000 to 30,000 years before present (B.P.). There are no younger Neanderthal fossils anywhere.

Origin of Modern Humans

Interestingly, anatomically modern humans and Neanderthals both have a lengthy chronological distribution. A skull from Petralona, Italy, for example, which may be about 200,000 years old, shows distinct Neanderthal tendencies but is not a classic Neanderthal (**Fig. 11.19**). On the other hand, fully modern humans are known from a site in South Africa known as Klasies River Mouth, which may be around 130,000 years old. Perhaps the most fascinating information regarding possible origins and relationships of modern humans and Neanderthals has become available recently from an area of the Middle East known as the Levant, a corridor region in Israel that hominids would be likely to take if leaving Africa. New electron spin resonance dates

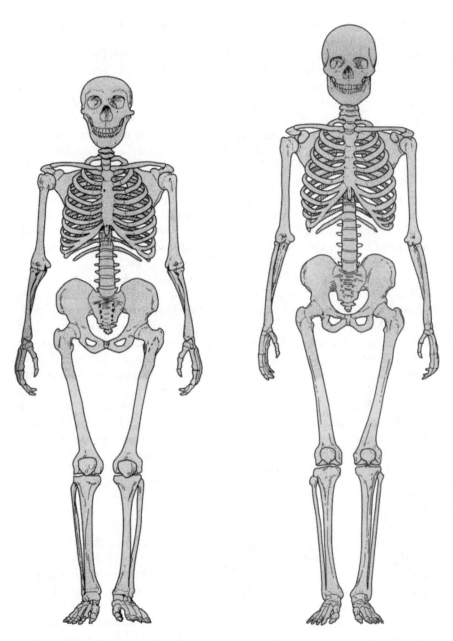

Figure 11.17 Comparison between a Neanderthal and modern skeleton. Neanderthals were relatively short with powerfully built bodies. As can be seen in this illustration, the brow ridges were more massive and the cranium lower in the Neanderthal. Many anthropologists now believe the Neanderthals were a different species from *H. sapiens*. (From B.G. Campbell, and J.D. Loy, 1995. *Humankind Emerging,* 8th ed. New York: Allyn and Bacon, p. 463.)

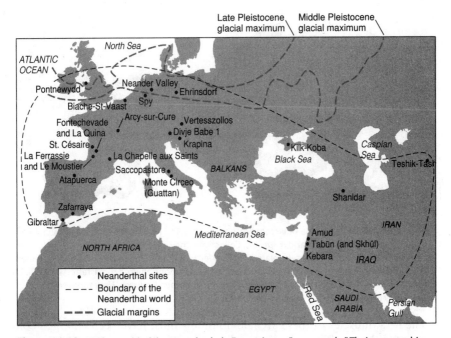

Figure 11.18 18 The world of the Neanderthals, Europe's true "cavepeople." Their geographic distribution seems to have been limited by the presence of caves satisfactory for their cultural needs. (Adapted and modified from C. Stringer and C. Gamble, 1993. *In Search of the Neanderthals.* London: Thames and Hudson.)

place anatomically modern humans in the cave of Qafzeh, Israel between 90,000 and 100,000 years B.P., about 50,000 years earlier than the Neanderthals that occupied nearby Tabun Cave and Shanidar Cave in Iraq. These are among the earliest records that express skull features we can label as distinctly Neanderthal and human, and if the dates are correct, then we can conclude that anatomically modern humans and Neanderthals overlapped in time for thousands of years, at least in the Middle East (**Fig. 11.20**). Clearly, both species in Europe and the Middle East came out of the morass of populations that paleoanthropologists have labeled as *H. heidelbergensis* and *H. antecessor*. If we believe the Spanish investigators, both evolved from the Atapuerca *H. antecessor* subsequent to about 0.78 Ma. This is certainly possible, and the age of the deposit is consistent with that interpretation. But the place and time for the earliest appearance of anatomically modern humans remains elusive. At this point all we can say is that there are no completely modern human skulls earlier than about 130,000 years B.P.

A variety of explanations for the appearance of the modern human skull exists in the literature. The heavy jaws of *H. ergaster* and its immediate descendants are combined with a somewhat enlarged occipital area (back of the skull), resulting in what is often called an occipital "bun" or torus (ridge).

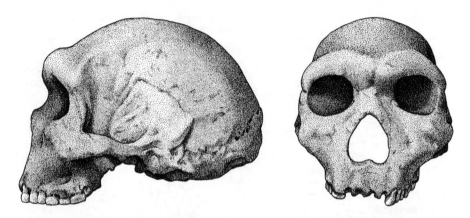

Figure 11.19 The Petralona skull from northern Greece. The Petralona people combined skull features of earlier *H. ergaster* and *H. sapiens*. Collectively they have been known as *H. heidelbergensis,* but some may be more closely related to Neanderthals than to modern humans, and it remains for future study to tease apart the phylogenetic patterns among them. (From I. Tattersall, 1995. *The Fossil Trail: How We Know What We Think We Know about Human Evolution.* New York: Oxford University Press, p. 175.)

The neck muscles of these archaic hominins must have been enlarged compared with ours. The reduction in these features, leading to a flatter face, reduced jaws, and smooth occipital region, led to the modern human skull shape. A number of authors have suggested that these changes were related to diet. Loring Brace, an American anthropologist, has proposed that the front teeth in archaic hominins, especially in Neanderthals, indicates they were used in many aspects of daily life, from eating to preparing animal hides. Improvements in foraging technology (especially in stone tools and eventually the discovery of fire) may have led to less use of the teeth and eventually to our modern skull shape. However, other scientists, especially the American David Pilbeam of Yale University, do not believe this argument can account for the changes. Instead, Pilbeam has suggested that the skull became modern in form to accommodate changes in the pharynx (throat area) accompanying the evolution toward the modern human vocal tract. He has pointed out that the modern human skull shape can be explained by a relatively simple developmental change resulting in a concave arch in the base of the skull that allows for the modern pharynx. Daniel Lieberman of Rutgers University has found that this change very likely was initiated by a shortening of a single bone, the sphenoid, in the base of the skull. The beauty of the Pilbeam hypothesis is that a mechanism of natural selection can be clearly envisioned; those individuals with greater verbal communication skills would have greater power in their communities and therefore would be most likely to mate. They would, in the biologist's vocabulary, have had greater fitness.

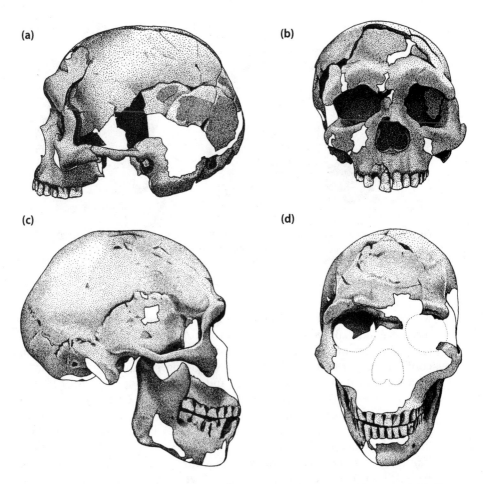

Figure 11.20 Archaic *Homo sapiens* from the Middle East. (a) and (b) side and front views of skull from Jebel Qafzeh, Israel; (c) and (d) skull from Skhūl, Israel. (From I. Tattersall, 1995. *The Fossil Trail: How We Know What We Think We Know about Human Evolution*. New York: Oxford University Press, p. 86.)

The first confirmed discovery of anatomically modern humans in association with extinct animals was made by railway workers as they were cutting a tunnel through a hillside in the Dordogne region of France in 1868 (**Fig. 11.21**). The workers unearthed bones and stone tools from a rock shelter known as Cro-Magnon, a name that has come to represent all the anatomically modern humans who inhabited Europe during the period from about 40,000 to 10,000 years B.P., a time also known to anthropologists as the Upper Paleolithic. Modern humans are unknown in Europe before this time, although as noted above their remains have been found in earlier sites in the

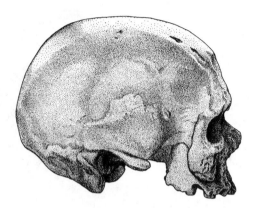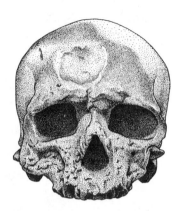

Figure 11.21 The famous Cro-Magnon "Old Man" from southwestern France; about 30,000 years B.P. (From I. Tattersall, 1995. *The Fossil Trail: How We Know What We Think We Know about Human Evolution.* New York: Oxford University Press, p. 25.)

Middle East and Africa. Neanderthals disappeared soon after modern humans began to spread throughout Europe. According to Bernard Campbell and James Loy in their outstanding text, *Humankind Emerging*, there are no confirmed records of *H. erectus* in Asia younger than about 0.30 Ma. So what happened to the Neanderthals and to ancient populations of *H. erectus, H. heidelbergensis,* and *H. antecessor?* To some extent, it depends on your philosophy of classification and evolutionary mode.

Many years ago an anthropologist named Carleton Coon proposed that modern human races first appeared in humans at the *H. erectus* grade of evolution. That is, our races were *older* than our species. His hypothesis, which became known as the "line and grade" model, suggested that *H. erectus* had dispersed from Africa throughout the Old World and had become adapted to regional environmental conditions. The characteristics that define our races, according to Coon, were features that conferred adaptive benefits to different habitats and climates and evolved separately in different world areas, whereas the populations that expressed them maintained some genetic continuity. *Homo sapiens* had not speciated from *H. erectus,* it had gradually evolved from it all over the Old World. Moreover, said Coon, the races did not need to have all passed the *H. sapiens* "threshold" simultaneously. And this was where he got into trouble, because he suggested that African blacks might have passed this threshold later than other races. Recent evidence suggests exactly the opposite, but there was no evidence at the time that Coon's book, *Origin of Human Races,* was published in 1962. He was branded as a racist among some circles, and even the label for his hypothesis, line and grade theory, was abandoned. In the current literature, exactly the same model is known as the "regional continuity," "multiregional," or "candelabra" model, and nobody pays homage to Coon for his initial research. I did not know the man, although I corresponded with him before his death. I thought his model was eminently plausible, and in 1970 I published one of the few scientific papers based on fossil evidence (in North American cotton rats, not humans) showing that the idea was at least feasible. Today we would likely use either *H. heidelbergensis* or *H. antecessor* as the direct ancestor of *H. sapiens* instead of *H. erectus,* but the model would otherwise remain the same.

Line and grade theory is championed now only by a few anthropologists such as the Australian Alan Thorne and Millard Wolpoff at the University of Michigan. Thorne, in particular, points to a number of fossil samples from Australia leading to living Australian aboriginal people that he believes supports the line and grade concept. In part, line and grade theory is out of favor because most systematists working with fossil and modern organisms prefer the punctuated equilibrium evolutionary model of evolutionary change over that of phyletic gradualism (see Chapter 2). Most younger systematists also prefer a cladistic approach to their taxonomy, and cladistics basically assumes a reticulate relationship among species in a clade. The jury is still out on the prevalence of punctuation version gradualism in the evolution of clades, and both have been identified in the fossil record (see the book *Morphological Change in Quaternary Mammals of North America*, edited by R. Martin and A. Barnosky, for some examples of phyletic change).

Nevertheless, despite the claims by Wolpoff and Thorne, additional studies, for example of the Upper Cave skulls from Zoukoudien, China dated at about 25,000 years B.P., fail to show the similarities to modern Oriental people that the multiregional hypothesis demands. Chris Stringer of the Natural History Museum in London found that these skulls are more similar to Cro-Magnon skulls from France (ca. 30,000 years B.P.) than they are to skulls of modern people from Japan (presumably similar to those from China). Also, the evidence for early fully modern people anywhere other than Africa is simply nonexistent, whereas it is present in African fossil localities. And this leads us, inexorably, to the "out of Africa" model for the evolution of human races.

The out of Africa (or "rapid replacement") model, proposed independently by a number of investigators, suggests that human races originated very recently, perhaps less than 100,000 years ago. Instead of being the latest humans to evolve, in this scenario African blacks are the oldest modern humans and gave rise to all other races. Changes in skin color and other features developed through natural selection as specific adaptations to regional environments, much as Coon had originally proposed. Humans probably exterminated other hominins along the way. Perhaps they ate them as well as just killing them. This is not as crazy as it sounds. There is evidence for cannibalism in *H. antecessor* at 0.78 Ma. It may be that the secret to our evolutionary success is the bizarre combination of intelligence and brutality. One could hardly argue against that based on occurrences even in the past century. The out of Africa hypothesis has gained some support lately in the work by Mark Stoneking and others, who have found, using mitochondrial DNA, the greatest amount of genetic variation in African blacks. His calculation of the origin for the ultimate Earth mother, or "mitochondrial Eve" as she is known in the literature, is about 130,000 to 140,000 years ago, which just happens to be consistent with the early record of modern humans from Klasies River Mouth Cave in South Africa. The rest, as they say, is history.

In a classic book published many years ago, the British anthropologist Desmond Morris discussed the close relationship between modern humans and the living great apes. He even named the book *The Naked Ape*, referring to the obvious hairless condition of modern humans. But when and under what circumstances would hairlessness have appeared? The only naturally occurring naked mammal is the naked mole rat of Northern Africa, *Heterocephalus glaber*. In addition to lacking fur and living underground, the naked mole rat is unique among mammals in that it is also the only species that demonstrates eusociality, that is, the same social structure as ants, termites, and bees. A "queen" mole rat does all the breeding in its colony and is attended by "workers." It seems quite clear that nakedness in humans did not arise for any of the same reasons as in mole rats. So what could have triggered this appearance? Although many adaptive scenarios abound, relating to thermoregulatory abilities of humans in humid tropical Africa, I suspect that as in the case of the evolution of limbs in tetrapods we should look to sex for the answer. Naked or seminaked skin appears regularly as a mutant condition in inbred strains of laboratory animals, especially mice and rats. It is not difficult to imagine a scenario in which the seminaked condition appeared in a hominin family and was considered attractive. It could also have been associated with enhanced intelligence or vocal ability. We know that animal features often are expressed during evolution in a mosaic fashion, so a combination of new features coupled with naked skin could have collectively increased fitness and thereby spread in the population. Of course, whether males or females were doing the selecting cannot be determined at this point. Although the natural tendency is to propose that males found naked females alluring, it could also be that a smart naked group of females selected which males they wanted for mates. Both are forms of "sexual selection" in biology, a well-known phenomenon that seems to be responsible for most, if not all, of the gaudy anatomical displays of male vertebrates (e.g., antlers and horns, peacock feathers, etc.). When humans got naked is anybody's guess. So far, we have no way to determine that from fossil materials.

Putting It All Together: A Summary of the Fossil Evidence for Human Evolution

The fossils from Chad and Ethiopia, *Sahelanthropus* and *Orrorin,* tell us that the ape–human split occurred before 6.0 Ma. Associated fossils and geological information also suggest that the environment of *Sahelanthropus* was forested and that the origin of bipedality therefore did not require our ancestors to first take to the savannas. Both apes and bipedal apemen walked the forests of Africa simultaneously. A later bipedal forest species, *Ardipithecus ramidus,* also combines characters of apes and humans, but its relationship to the earlier *Orrorin* and *Sahelanthropus* and to the later *Kenyanthropus* and *Australopithecus* remains undetermined. *Sahelanthropus* and *Kenyanthropus* have decidedly flattened faces, which may or may not indicate a phylogenetic alliance. These species represent part of a "bush" of radiating ape–human experiments (**Fig. 11.22**), all of which were partly or completely bipedal, and more are likely to be discovered. This is consistent with results from studies of nonhuman mammal clades, such as the rodents, in which a mosaic evolutionary response is common during the early radiative period (see discussion of mosaic evolution in Chapter 2). It is as if a pool of characteristics is shared almost at random among a series of related species. In this early anatomical experimental phase, which Alexey Tesakov (Russian Academy of Sciences) and I have labeled the "metaregion" of clade morphospace (the range of anatomical possibilities), it is generally difficult or impossible to determine which species are ancestral to later groups. So it should not be surprising that paleontologists are coming up with a host of early hominid species that are obviously somehow related to each other but in ways that are difficult to determine. With rodents, we often have hundreds of specimens, but with the few hominid specimens available it is little wonder that the story takes so long to unfold.

The next segment of our family tree is represented by a series of austraopithecines. After a certain point, they are easily recognizable as the gracile *Australopithecus* and robust *Paranthropus* groups. Although broken into three species, *Paranthropus aethiopicus* may be an ancestral form of the later eastern *P. boisei,* and *P. robustus* may represent just the southern populations of the same species. So only one robust species might have been around at any given time. Relationships among the species of *Australopithecus* are more difficult to determine. *Australopithecus anamensis* is the earliest, but its relationships to *Kenyanthropus* and other *Australopithecus* species is obscure. *Kenyanthropus* presents a particularly vexing problem, as it was contemporaneous with *A. afarensis* and has a flatter face. Perhaps *A. afarensis* was ancestral, not to later species on the line to modern humans but only to the *Paranthropus* group.

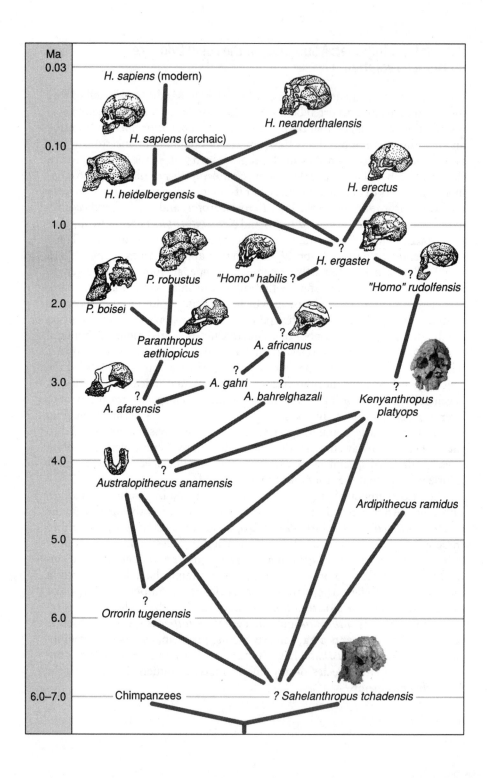

Then there is the problem of *H. habilis* and *H. rudolfensis*. With relatively large brains and flat faces, they are the earliest members of our own genus. Their immediate ancestors have not been determined, perhaps because they do not exist among the known hominids. One of the striking features of *H. habilis* is its small size. The immediate ancestor of later African hominids such as *H. ergaster* must have been much larger, so *H. habilis* cannot be the direct ancestor of later more advanced species. The skull of its contemporary, *H. rudolfensis*, sometimes considered synonymous with *H. habilis*, is a bit larger, but there still remains both a size and morphological gap between it and *H. ergaster*. Meave Leakey has pointed out the facial similarities between *K. platyops* and *H. rudolfensis*, so this could represent a possible tie.

The Nariokotome boy *H. ergaster* skeleton is essentially modern below the neck, but the skull has many primitive features showing alliance with European *H. heidelbergensis* and southeast Asian *H. erectus*. Its ancestry is unclear, though it was certainly derived from a species with some of the features of *H. habilis* and *H. rudolfensis*. *Homo ergaster* was the first species of our genus to leave Africa, probably between 1.7 and 1.8 Ma, which corresponds worldwide to the beginning of the Pleistocene Period, or the Ice Ages, when sea levels were considerably depressed and various continents and islands were newly connected as a result. Continental glaciers extended down into the mid-latitudes, and life in general would have been very challenging in the north. But a large mammal megafauna, composed of mammoths and other large ungulates, might have made easy prey for an intelligent and highly social hominid species.

After dispersal from Africa, humans continued to diversify anatomically into at least three species that we recognize at the end points of their lineages as the southeast Asian *H. erectus,* the European Neanderthals (*H. neanderthalensis*), and modern humans (*H. sapiens*). *Homo heidelbergensis* and *H. antecessor* represent populations that are morphologically and chronologically intermediate between *H. ergaster* and both Neanderthals and modern humans. It is not at all clear if they are distinct species or just populations whose relationships we cannot yet determine. In one evolutionary scenario, the early *H. ergaster* emigrant from Africa (represented in the fossil record at Dmanisi, Georgia) was not ancestral to modern humans and gave rise instead only to southeast Asian *H. erectus*. In this scenario, there may have been a series of human dispersals out of Africa, one leading through *H. an-*

Figure 11.22 22 A summary phylogeny for hominids. Despite many new exciting discoveries in the past few years, the relationships among the species remains difficult to determine. This is not an argument against the evolution of modern humans from these ancient people, rather it simply reflects the difficulty in finding hominid skulls. Albeit a bit complicated, this illustration represents our current best estimate of the human pedigree, or family tree. (Images from I. Tattersall, 1995. *The Fossil Trail: How We Know What We Think We Know about Human Evolution.* New York: Oxford University Press, the Mission Paleoanthropolgique Franco-Tchadienne, and National Museums of Kenya.)

tecessor to the Neanderthals and a much later one, perhaps less than 100,000 years ago, leading to modern human races. This is the out of Africa hypothesis for the origin of races. The line and grade or candelabra model suggests instead that we are all descended from an *ergaster-* or *erectus*-grade hominid and that the evolution of human races appeared early in our history, actually before the appearance of a fully modern skull.

The fossil record shows that fully modern humans were around by about 100,000 years ago, and that makes us contemporaries with the Neanderthals, who dominated Europe through the Pleistocene. Our ancestors may have even shared valleys in the Middle East. But Neanderthals were gone by around 30,000 years B.P., when we see the famous Cro-Magnon human populations in Europe. My bet is that the prime human characteristics of intelligence, cruelty, and greed won the day. Just watch the stock market.

Suggested Readings

Asfaw, B., T. White, O. Lovejoy, B. Latimer, S. Simpson, and G. Suwa. *Australopithecus garhi,* a new species of early hominid from Ethiopia. *Science* 284:629–635.

Bermúdez de Castro, J. M., J. L. Arsuaga, E. Carbonell, A. Rosas, I. Martínez, and M. Mosquera. 1997. A hominid from the lower Pleistocene of Atapuerca, Spain: possible ancestor to Neanderthals and modern humans. *Science* 276:1392–1394.

Brunet, M., A. Beauvilain, Y. Coppens, E. Heintz, A. H. E. Moutaye, and D. Pilbeam. 1996. *Australopithecus bahrelghazali,* une nouvelle espèce d'Hominidé ancien de la région de Koro Toro (Tchad). *Comptes Rendus* 322:907–913.

Brunet, M. et al. 2002. A new hominid from the Upper Miocene of Chad, Central Africa. *Nature* 418:145–151.

Caan, R. L., M. Stoneking, and A. C. Wilson. 1987. Mitochondrial DNA and human evolution. *Nature* 325:31–36.

Campbell, B. G. and J. D. Loy. 2000. *Humankind Emerging* (eighth edition). Allyn and Bacon, Boston.

Goodall, J. 1986. *The Chimpanzees of Gombe.* Harvard Unviersity Press, Boston.

Gore, R. 1996. The neandertals. *National Geographic* 189:2-35.

Grine, F. E. 1988. *The Evolutionary History of the Robust Australo-pithecines.* Aldine, New York.

Gunnell, G. F. and E. R. Miller. 2000. Origin of the Anthropoidea: dental evidence and recognition of early anthropoids in the fossil record, with comments on the Asian anthropoid radiation. *American Journal of Physical Anthropology* 114:177–191.

Johanson, D. C. and M. A. Edey. 1981. *Lucy: The Beginnings of Humankind.* Simon & Schuster, New York.

Leakey, M. G., C. S. Feibelo, I. McDougall, and A. C. Walker. 1995. New four-million-year old hominid species from Kanapoi and Allia Bay, Kenya. *Nature* 376:565–571.

Leakey, R. and R. Lewin. 1992. *Origins Reconsidered: In Search of What Makes Us Human.* Doubleday, New York.

Lemonick, M. D. and A. Dorfman. 2001. One giant step for mankind. *Time* (July):54–61.

Lewin, R. 1998. *Human Evolution.* Blackwell Science, Malden, MA.

Martin, R. A. 1981. On extinct hominid population densities. *Journal of Human Evolution* 10:427–428.

Senut, B. et al.. 2001. First hominid from the Miocene (Lukeino Formation, Kenya). *Comptes Rendus* 332:137–144.

Semaw, S., P. Renne, J. W. K. Harris, C. S. Feibel, R. L. Bernor, N. Fesseha, and K. Mowbray. 1997. 2.5-million-year-old stone tools from Gona, Ethiopia. *Nature* 385:333–336.

Simons, E. L. and T. Rasmussen. 1998. A whole new world of ancestors: Eocene anthropoideans from Africa. *Evolutionary Anthropology* 3:128–139.

Stringer, C. and P. Andrews. 1988. Genetic and fossil evidence for the origin of modern humans. *Science* 239:1263–1268.

Tattersall, I. 1995. *The Fossil Trail.* Oxford University Press, New York.

White, T. D., G. Suwa, and B. Asfaw. 1994. *Australopithecus ramidus,* a new species of early hominid from Aaramis, Ethiopia. *Nature* 371:306–312.

Wolpoff, M. and R. Caspari. 1997. *Race and Human Evolution.* Simon & Schuster, New York.

Wright, K. 1999. First Americans. *Discover* (February):53–63.

12

Yesterday, Today, and Tomorrow

"Indeed, evolution is happening today, sometimes at a blistering rate that leaves evolutionary biologists in awe."

The title of this chapter derives from the film by the same name, with the late great actor Marcello Mastroianni and the beautiful and talented Sophia Loren. The movie includes three vignettes, taking place at three different periods in the lives of the protagonists. Just so, evolution has a past, present, and future. The past is only accessible through the fossil record, and the primary purpose of this book has been to use the fossil record to document the evolutionary process, to show that missing links abound. As we have seen, historical links can be seen in the fossil record on a variety of time scales, from a few thousand years of the Pleistocene to the multibillion year history of life on Earth. But what can we say about links in modern time? Have organisms changed at all during historical time? Is there any possibility that today we are witnessing changes in populations that may lead to new species? Is it possible to experiment with the evolutionary process, and in doing so illuminate how fast change can take place? Domestic breeders have known for thousands of years that artificial selection by humans can dramatically change organismal form; after all, that is how we get corn, Idaho potatoes, 10-pound tomatoes, Jersey cows, and the dachshund. In the laboratory, geneticists have bred selected lines of the fruitfly, *Drosophila melanogaster,* with abdominal bristle numbers varying from 3 to 85 and mice that varied in weight by more than 300% in less than 50 generations. Reproductive isolation, the hallmark of speciation, has been achieved in laboratory *Drosophila* in less than 20 generations. Humans may also have unwittingly induced speciation in the wild. New species of annual grasses may be evolving to grow on strip mine soils heavy in copper. Indeed, evolution is happening today, sometimes at a blistering rate that leaves evolutionary biologists in awe. The examples in this chapter deal with evolution in wild populations occurring today and in the recent past. To set the scene for the investigations that will deal solely with evolution and selection during our lifetime, let us look at another example of rapid evolution during the late Pliocene and Pleistocene, this time with fresh-water snails.

A Snail's Pace Is Not Necessarily Slow

The region east of Lake Turkana in the Great Rift Valley of East Africa has become famous for producing some of the most important hominid fossils in the world. In particular are the enigmatic skulls variously referred to *Homo habilis* and *Homo rudolfensis*. Less well known are the other animal remains from this area. As in all fossil-rich sediments, many groups are often represented, but in this case they have been overshadowed by the hominid finds. Not too long ago, however, an English paleontologist, P. G. Williamson, made quite a splash by reporting rapid evolutionary change in 13 lineages of fresh-water mollusks from about five million years of lacustrine (lake) sediments that accumulated near Lake Turkana (**Fig.12.1**). Williamson took many measurements from a variety of snails and bivalves throughout the sequence and then compared them all to determine if any changes had occurred and where in the sequence the changes were found. What makes this study unique is the relatively complete record Williamson had to work with. There were literally thousands of snails from 190 separate fossil localities. His overall finding was that the snails did not change for long periods of time, and when they did evolve they changed rapidly and often simultaneously. In addition to shape changes within lineages, Williamson claims to have recorded the rapid origin of new species. All the directional changes appear to have taken place over periods of time between 5000 and 50,000 years. Although perhaps not rapid change to a geneticist, the evolutionary shifts in shell shape took place over only a brief period of the full history of each species lineage, and so the change conforms to Eldredge's and Gould's model of punctuated equilibrium rather than to the standard Darwinian model of phyletic change, the latter predicting some kind of change throughout much of the history of each lineage. It will be worthwhile to keep these results in mind as we look now at some examples of evolution during the lifetime of a single observer.

Evolutionary Change Among Living Animals

Founding Voles

Clethrionomys glareolus is a small common wild rodent in Europe; in Great Britain it is called the "bank vole." It is another member of that group of ubiquitous, north temperate, arvicolid rodents. Like arvicolids everywhere, its

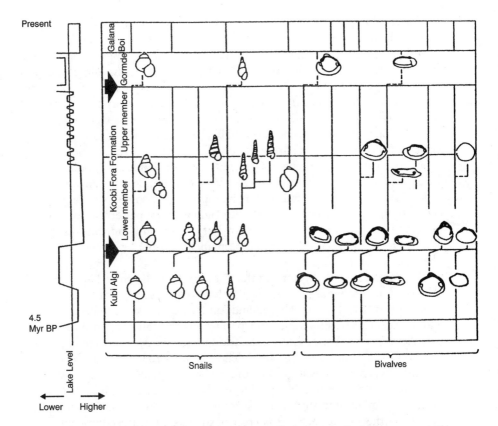

Figure 12.1 A record of late Pliocene and Pleistocene mollusks from the Koobi Fora area, east Africa, over the past 4.5 million years. P. G. Williamson interpreted this history to reflect an example of punctuated equilibrium and the rapid origin of new species. However, mollusks are notorious in changing form under varying environmental conditions, so some of the "species" could be ecophenotypes and, thus, change within lineages could be phyletic after all. (From P.G. Williamson, et al., 1981. Paleontological documentation of speciation of Cenozoic molluscs from the Turkana Basin. *Nature* 293: 437–433. copyright Nature Publishing Group.)

numbers may fluctuate greatly, with times of immense population explosions followed by times of considerable scarcity. These cycles have been studied by ecologists for many years and seem to be controlled predominantly by available forage; the rodents reproduce to the point that they use up the available food and then the population crashes. The cycle may take many years to play out, and predator numbers often oscillate in sympathy. The British mammalogist G. B. Corbet examined collections of the bank vole made over a number of years when it colonized a set of young forest plantations near Lock Tay, Perthshire in Scotland. He recorded an interesting temporary change in dental morphology that may give us a clue as to how founder effect works(see Chapter 2).

The third upper molar of *C. glareolus* displays two morphological states, or morphotypes: the simplex and complex condition. In the simplex condition there is one less triangle than in the complex form. In collections made in 1955 and 1957, Corbet found that populations from one plantation (Borland) were unique in their expression of the complex morph. Another nearby plantation (Balnearn) showed only the simplex morph, more typical also of mainland Scotland. By 1972, when the populations were again sampled, the Borland animals had reverted back to the simplex state. Corbet was unsure if the return to the ancestral condition was due to selection against complex morphs or just genetic swamping from the Balnearn population, but he seemed convinced that the prevalence of the complex morph was due to colonization by animals with a higher proportion of complex M3s, which is a true founder event. The return to the original dental pattern shows that the observed change was only temporary, but is there any indication that this kind of mechanism could lead to a permanent morphological shift? In fact, there is.

A similar dental dichotomy of the third upper molar shows up in the fossil record in another arvicolid, the North American sagebrush vole, *Lagurus curtatus* (**Fig. 12.2**). John Rensberger and Anthony Barnosky documented a gradual replacement of the simplex morphology by the complex form during the late Pleistocene in Washington state. Modern sagebrush vole populations exhibit only the complex morphology. The locality, a road cut near the town of Kennewick, spans at least 40,000 and perhaps as much as 328,000 years of the latest Pleistocene and early Holocene. Simultaneously with changes in the M3, the number of triangles increases on the first lower molar, from four to five. Over the long term, the fossil record shows that dental complexity in arvicolids is generally favored, probably in part because complexity is also often linked with body size increase. Corbet's study, over a very brief period of time (an insignificant tick on the geological clock), gives us a clue to the kind of variation expected and the mechanisms involved in establishing an anatomical variant in ecological time, the lifetime of an individual human observer.

Of Blue and Brown Adders

The "new synthesis" of evolution, combining modern knowledge of heredity with Darwinian natural selection, suggests that evolutionary changes can come about in populations with the spontaneous origin of a nonlethal mutation that confers a reproductive advantage to those individuals that possess it. In the laboratory, many such systems have been artificially constructed, but in nature, where change occurs more mysteriously and with less regularity, it is not easy to document. So I was very pleased and excited to see that a case with all the necessary ingredients had been observed in a population of snakes from southern Sweden.

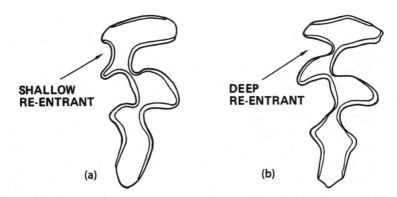

Figure 12.2 The simplex (a) and complex (b) morphotypes in the third upper molar of the extant North American sagebrush vole, *Lagurus curtatus*. Over a period of thousands of years during the Pleistocene in Washington state, the complex form replaced the simplex type, and only the complex form is found in living populations. (From R.A. Martin, and A.D. Barnosky, eds., 1993. *Morphological Change in Quaternary Mammals of North America.* New York: Cambridge University Press. p. 335.)

In a paper published in the journal *Evolution* in 1992, Thomas Madsen and Richard Shine reported an example of microevolutionary shifts in coloration, body size, and sexual size dimorphism across generations in a poisonous adder, *Viperus berus*. Adders mate in southern Sweden in the spring, and females give birth during the late summer. Males mature in 3 to 4 years, and females breed for the first time after about 4 years. Females generally attain larger size than males, and reproductive success increases with body size in both sexes. Larger females produce larger litters than smaller ones, and larger males win more combat bouts, thus inseminating more females.

Madsen and Shine studied their adders for 10 years. In 1983 they captured an adult female with a unique color pattern; it was gray–blue dorsally with a dark gray mid-dorsal zigzag line. Adders are normally light brown, with a dark brown zigzag line. In subsequent years they found more blue snakes, with the males in particular more spectacularly colored, pale "sky blue" with a jet-black mid-dorsal line. The blue colors do not appear in both sexes until sexual maturity, so it is clear the allele (or alleles) for this color were already present in the population 3 years before discovery of the first blue snake. Because females mate more than once and litters are sired by more than one male adder, it has not been possible yet from field work to determine the genetic structure of the population. Nevertheless, some interesting results have been observed since 1983.

Blue and brown females appear not to differ in most measurable quantities, but there has been a slight increase in female body size of both color morphs over the study period. For males the changes have been dramatic. Blue males are larger, mate more frequently, and have faster growth rates than their brown relatives. The allele(s) coding for blue color has resulted in the size of males approaching those of females, reducing the sexual dichotomy from 13% to 10% (remember, adult females are normally larger than males). The number of blue individuals has also risen, both in males and females; the changes appear to be favorable and are spreading. The authors demonstrated that the spread of the blue morph and its controlling allele(s) is the result of selection primarily on the males; the blue males have increased "fitness," because they breed more effectively than the brown males, which they are quickly replacing. This kind of evolutionary change is known as sexual selection, and is believed to be a powerful force in evolution, resulting in many of the differences we see among the sexes in wild animals. Examples include different body size between males and females in many vertebrates and other secondary sexual characteristics such as horns, antlers, and tusks in some mammals; gaudy feathers in birds; long tails and breeding coloration in male fishes; and so on. It is very satisfying to finally see a natural example of this phenomenon, and it reinforces the conclusion that we need to continue long-term studies of natural populations to give ourselves the time necessary to at least see the fringe of evolution in action.

The Snails of Galicia

The broad panorama of snail evolution Williamson gave us for Lake Turkana is at a scale that disallows identification of the mechanisms involved in speciation, the origin of new species. For that, we either need to have many more fossil localities represented in the refined history of some organism or we need to see the process in action among living organisms. Either will likely be a rare find, because new species seem to come into existence rapidly, probably derived from very localized ancestral populations. Nevertheless, many "experiments" are likely going on in modern populations, some of which may result in new species and some not. In Mayr's Biological Species Concept, the salient point is whether or not two populations interbreed and share their genes. If they do, then they are not yet different species (with the exception of occasional hybridization, which can be expected from time to time). So, it may be possible to find populations of living species that seem to demonstrate at least some incipient block to reproduction. Although not proof positive of speciation in action, such discoveries would certainly demonstrate that the potential for the origin of new species is present. Three biologists, Kerstin Johannesson, Emilio Rolan-Alvarez, and Anette Ekendahl, recently described such a situation in a shallow water marine snail, *Littorina saxatilis*, in Galicia, Spain.

Littorina saxatilis is a highly polymorphic snail found in the rocky littoral zone, the area between high and low tides. Gene flow is slow between breeding populations, and there is occasional transport of individuals between local islands. Adults move on the order of 1 to 3 meters every 3 months. The snail has separate sexes and internal fertilization. Because there is only minimal gene flow between local populations, different shell morphologies show up in local areas, sometimes even at different levels of the same shoreline. In Galicia, a smooth unbanded form is confined to a low shore level of blue mussels. Ten to 20 meters higher up the shore a ridged and banded form is found in a level of barnacles. The forms overlap in a narrow zone on the border of the mussel–barnacle levels. Hybrid individuals, with intermediate shell morphologies, are found in the overlap zone at an average of about 18% of all snails observed.

The authors found that the snails do not mate at random where they overlap. This seems to be due to two factors: a real preference for snail morphs of their own type and a tendency to run into more snails of their own type because they prefer slightly different habitats, even within the overlap zone. In two areas sampled, only 1 of 60 and 8 of 34 pairs of mating snails were of different morphs, showing conclusively the preference of morphs to mate among their own kind. But what happens to the hybrids? Are they as "fit" as either adult morph? The authors have no long-term data analyzing this. However, it is clear that the hybrids are uncommon and, at least during the time of the study, have not replaced the two dominant morphs. This suggests that the hybrid zone is stable and, further, that natural selection is operating to reduce gene flow between the two morphs. Whether we are seeing the origin of a new species or just a temporary displacement in morphology remains for another generation of dedicated biologists to determine.

All House Sparrows Are Not the Same

It is a characteristic of human nature, sometimes lamentable, to modify the natural landscape. Homes are built, land is cleared, gardens are planted, factories go up. As if that is not enough, exotic species must be introduced. America abounds in numerous animal species that did not originate here; walking catfish, European wild boars, starlings, oscars (a cichlid fish), and house sparrows are but a few. Some, such as the starling and house sparrow, have shown an amazing propensity to adapt to their new environment and have dispersed quite literally throughout North America. A number of studies have shown that the house sparrow has diverged anatomically since its original introduction in the United States; in 1971 Robert Johnston and R. K. Selander measured 16 skeletal size variables in populations of this sparrow and showed that by that time it had developed a continental pattern of size distributions, indicating adaptation to local environments. The sparrows were larger in the north than in the south, following what is commonly referred to as Bergmann's Rule. There are some exceptions to this pattern in some species, but it is a pattern seen in many animals. The idea is that the larger animals are better able to cope with severe northern winters by putting on enough fat stores to get by during inclement weather. As endotherms, or "warm-blooded" animals, larger individuals would also lose less heat per unit body area than small ones.

Darwin's Finches

Individual events can often have profound influences on people's lives—sometimes good, sometimes bad. When Charles Darwin, at the age of 22, joined the crew of *H.M.S. Beagle* in 1831 as the ship's naturalist, there was no way to predict that this single voyage would shape a young man's mind and change the course of history. The purpose of the trip was exploratory, much like the mission of the starship *Enterprise* in the "Star Trek" saga. Darwin collected examples of modern plants and animals as well as fossils. He interacted with primitive human populations, and he did a lot of reading and thinking. At one point in the voyage, in 1835, the Beagle stopped off the coast of Ecuador, South America to reconnoiter among the Galapagos Islands. It was here that Darwin became enthralled with organismal diversity, especially among the tortoises and the birds. The Galapagos had been known for some time. Sailors would stop there to take on food and water and to "harvest" the tortoises, because some of the species were huge, weighing hundreds of pounds. A human could ride on the largest. Darwin became especially interested in the ground finches, and his studies, as well as those of later biologists such as David Lack, have shown a considerable radiation of these birds in the Galapagos (**Fig. 12.3**). Finches there exist in a variety of sizes, and their beaks are designed for a range of diets, including seeds, cactus flesh, buds, leaves, fruits, and insects. One species is actually a tool user, picking insects out of holes in trees with a cactus spine or stick. It does not impale the insect; rather it uses the stick to induce the insect to emerge from the hole; it then drops the stick to eat the bug.

The Galapagos Islands are isolated, and most are barely accessible. These features appealed to a husband and wife team, Peter and Rosemary Grant of Princeton University, and they have spent much of their professional careers in this "natural laboratory," studying the finches. During this time, they have seen natural selection in action and given us another glimpse at how evolution works to both change and preserve morphologies. A period of time in the late 1970s was particularly instructive. The Grants were following both the weather and also a number of features of the seed-eating finch *Geospiza fortis* and its environment on the island of Daphne Major. They recorded changes in such quantities as beak size and the relative abundance and size of seeds the finches feed on. The Galapagos normally experience a cycle of weather that includes a hot wet season in January through May followed by

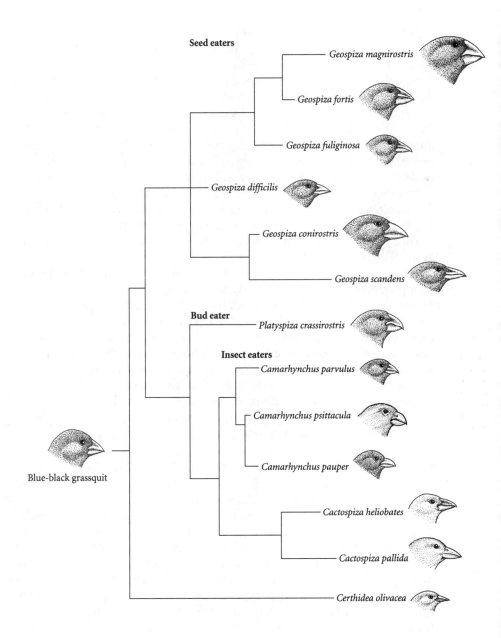

Figure 12.3 The diversity of finch species living today on the Galapagos Islands off the coast of Ecuador. This phylogeny, supported by a variety of morphological and genetic studies, suggests an origin from a generalized South American mainland finch. (From D.J. Futuyma, 1998. *Evolutionary Biology*, 3rd ed. Sunderland, MA: Sinauer Associates. p. 118.)

a cool dry season for the remainder of the year. In 1977 the rains never came, and there was almost a 2-year dry period. Most of the finches died; the females were harder hit than the males.

Before the drought hit, seeds were present in small through large sizes. As the drought persisted the plants produced, on the average, larger fruits. It turns out that larger finches, with thicker larger beaks, can obtain seeds from the fruits of native plants more efficiently than can the small ones. As the drought went on, the smaller finches began to die off, shifting the average (mean) size of the population up. Females died at a faster rate than males because they average slightly smaller to begin with. The Grants had already shown that size was inherited in finches, so the results of this natural selection resulted in an increase in average size in the next generation. Birds born in 1978 were about 24% larger than those born before the drought.

Four years later the weather reversed. Because of a periodic phenomenon in the Western Hemisphere known as "El Niño," the trade winds changed direction, and the dry season in the Galapagos became one of torrential rains. Seed production was extraordinary, and more small fruits were produced. Theoretically, this should have favored the small birds, and that is exactly what happened. Finches born in 1984 and 1985 had beaks about 2.5% smaller than those before the rains came. This study is also a fine example of how environmental influences act to maintain morphology within a particular range, what we have seen in section I as "stabilizing selection." Nevertheless, if the weather was to change dramatically in the Galapagos, toward either greater aridity or more humid conditions, we can see very clearly what forces would be in operation to link one kind of bird with another through time. The mechanisms of evolution, and of "missing links," are becoming accessible.

Sympatric Divergence in Salmon: You Are Where You Are

As we saw in Chapter 2, one hypothesis of speciation is called the allopatric model. In this scenario, a daughter population fragments from the parent population and after time diverges to the point that when the two populations again overlap, they no longer recognize each other as potential mates, or mating is inviable (offspring are not fertile, as in a mule, the offspring of a donkey and a horse). Another hypothesis, called the sympatric speciation model, proposes that new species can arise from within the geographical range of a parent species if the daughter population is ecologically separated. Sympatric speciation is recognized as a likely pathway for some insects and has been suggested for the pea aphids, fall armyworms, soapberry bugs (**Fig. 12.4**), larch budmoths, and other species. The apple maggot fly of North America presents another good example. Larvae of the ancestral form of this fly feed on hawthorn trees. After the introduction of apples from Europe to this country about 300 years ago, some maggot flies apparently began to feed

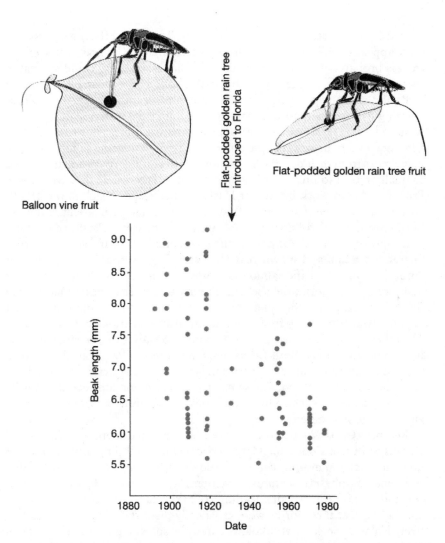

Figure 12.4 Change in body form of the soapberry bug as a function of diet. The normal food of these bugs is the fruit from the native balloon vine. In 1920 the flat-podded golden rain tree was introduced to Florida, with smaller fruits. Average beak lengths of soapberry bugs decreased over the next 40 years as they made greater use of this resource. (*Evolutionary Analysis,* 2e, by Freeman/Herron, copyright 2001. Reprinted by permission of Pearson Education, Inc., Upper Saddle River, NJ.)

and breed on them instead of hawthorns. Jeffrey Feder and colleagues have shown that the maggot flies breeding on these different tree species are genetically different. It is not too difficult to imagine that insects might be able to diverge in the same geographical range, because they are fairly small and limited in their distribution, even if they fly. To them, different tree species must appear almost as entirely different ecosystems. The grass between the trees is probably just like water between islands. But how could sympatric speciation occur in a large animal that swims hundreds of miles, like a salmon?

Pacific salmon spawning runs are known to just about everyone. These amazing fish are an important part of the commercial fisheries of the Pacific west and an integral part of the balance of nature along the eastern Pacific Rim. Each year, mature sockeye and other species of Pacific salmon migrate many miles from their oceanic haunts up into freshwater rivers and tributaries to finally return to breed in the same stream or lake where they were born. Primarily due to work by Arthur Hasler and colleagues back in the 1950s, we now understand that salmon find their way home with a remarkably refined sense of smell, following chemical clues in the water. Pacific salmon migrate only once; they die after spawning. Atlantic salmon and steelhead trout migrate upstream and spawn year after year. Two morphological forms (or "morphs") of the Pacific salmon *Oncorhychus nerka* have been known for many years. One form, the sockeye, spends a number of years maturing in the Pacific Ocean and then migrates into fresh water to spawn. Animals that migrate from saltwater to fresh water to breed are called *anadromous* species. The other form, the kokanee, grows up in deep waters of a freshwater lake. It is nonanadromous. Both forms make nests in shallow gravelly streams. The sockeye and kokanee are very similar in body characteristics and have been shown to interbreed. Consequently, they are not considered to represent distinct species. Then why do they look and act different from each other, and how did that difference begin?

Recently, Chris Wood and Chris Foote of the Department of Fisheries and Oceans in British Columbia, Canada decided to see if they could figure out how these salmon morphs came to be so distinct. They evaluated reproductive behavior, physical features, development, and genetic features in sockeye and kokanee in streams leading to Takla Lake, in British Columbia, concluding that these forms were genetically distinct and mostly reproductively isolated. Sockeye and kokanee morphs spawning in the same stream are much more genetically different from each other than either is from the same morph spawning in different streams. Kokanee have more gill rakers than sockeye, and this difference has been heritable for at least two decades. Gill rakers are cartilaginous extensions from the gills that aid in straining food from the water. Sockeye grow more rapidly and longer than kokanee, which seems to be an advantage for a form undergoing the rigors of a seagoing migration. Most kokanee and sockeye mature at 4 years of age, but they are of different size at this age, and differential mating occurs according to size. Kokanee and sockeye mate almost exclusively with their own morph. The smaller size of adult kokanee also appears to be adaptive; they do not have the same large forage base in Takla Lake as the sockeye will have in the ocean. Hybrids are formed between kokanee and sockeye only by male kokanee "sneak mating," where kokanee males will mate with both sockeye male and female pairs and lone females. In one creek, just over 20% of sockeye eggs per nest were fertilized by kokanee males. However, Wood and Foot estimated only about 0.8% gene flow between morphs spawning in the same tributary. Although they have no conclusive data at this time, Wood and Foot surmise that selection against hybrids must be fairly high. Given the theoretical energy needs of the two morphs, this makes sense. Hybrids would not likely be large enough to survive the ocean-going trip, but they would be too large to survive and reproduce on the available forage in Tekla Lake. Also, in reduced numbers, where mating is assortative by size, they would have a reduced probability of finding appropriate mates.

So how did the sockeye and kokanee morphs appear, and what continues to promote their divergence? In some circumstances anadromous (migratory) sockeye produce nonanadromous offspring called "residual sockeye." These residual individuals are usually males and smaller than the anadromous individuals. As noted above, sockeye are known to mate assortatively by size. This would naturally lead to an initial population of smaller anadromous individuals, and natural selection would further refine the anadromous population, leading to the kokanee morph. Although sockeye and kokanee may spawn at the same time in streams, in Takla Lake kokanee spawn later and in shallower water than sockeye. This would act as an additional isolating mechanism leading to further divergence. Perhaps the most fascinating aspect of the origination of the kokanee from the sockeye morph is that it is not apparently a "monophyletic" occurrence. That is, all kokanee morphs are not descended from a single ancestral population. Rather, most kokanee populations seem to have been independently derived from separate parent sockeye populations.

If all sockeyes were to go extinct tomorrow, we would have the interesting circumstance where a remaining "species" was descended from multiple ancestral populations, each slightly genetically distinct from one another. The same pattern of origin has also been reported in the three-spined stickleback, a freshwater fish in western Canada that evolved from multiple populations of a marine relative (**Fig. 12.5**). I wonder just how frequently this happens in nature, and not only in fish.

Experiments in Natural Selection

Guppies and Predators
The lovely guppies we find in our tropical fish stores are also experimental animals in some fascinating research being conducted by evolutionary biol-

Figure 12.5 The multiple origin of populations of freshwater three-spined stickleback fishes from different populations of a marine ancestor. In (a) a large marine ancestor has given rise to smaller open-water forager forms (solid polygons) in various rivers. In (b), different populations from the same ancestor become established in lakes, where they later diverge independently into open-water foragers and bottom-foragers (hatched polygons). This "parallel speciation" in the making may also apply to Paciifc salmon. (From D.J. Futuyma, 1998. *Evolutionary Biology*, 3rd ed. Sunderland, MA: Sinauer Associates. p. 489; after D. Schluter, and L.M. Nagel, 1995. Parallel speciation by natural selection. *American Naturalist* 146:292–301.)

(a)

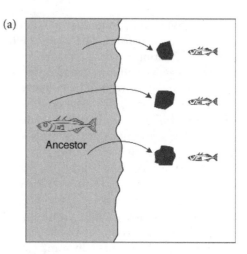

(b)

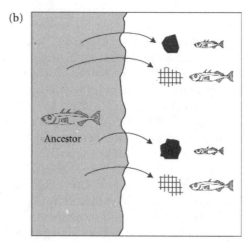

ogist David Reznick. He and his colleagues have been studying the natural history of guppy populations in their native streams in Trinidad. Depending on where they are located, the populations live in what Reznick calls either high or low predation communities. High predation communities occur wherever the guppies coexist with larger predatory fish of the Family Cichlidae, in particular the pike cichlid *Crenichla alta*. Parts of the streams are blocked by waterfalls, and in some pools the guppies are forage only for smaller, less voracious predators, such as the killifish *Rivulus hartii*. The cichlids prefer to dine on large adult guppies, whereas the killifish choose the smaller juvenile guppies. In a number of studies Reznick showed that the guppies respond to these predators with different reproductive strategies. When reared in the laboratory, guppies from high predation localities attained maturity earlier and at a smaller size than those from low predation sites. They also produced more litters and more offspring per litter than guppies from low predation locations. All these differences were shown to be heritable. However, these kind of studies, by definition, are *inferential,* in the sense that they do not represent changes that were followed by the investigator. To confirm the fact that natural selection by predation was in fact responsible for these differences and also to test the rate at which these differences could appear, an experiment was required.

Reznick and colleagues discovered locations on tributaries of two rivers, the El Cedro and Aripo, where high predation guppy populations existed below waterfalls; above the waterfalls guppies were absent and only the small guppy predator, the killifish *Rivulus hartii,* was present. So the research team transplanted some of the high predation guppies above the waterfalls. Evaluations were made in the Aripo River stream 11 years after the introduction and both 4 and 7.5 years after introduction in the El Cedro stream. The life histories of the Aripo transplants had evolved to be virtually identical to those of the naturally occurring low predation populations; the descendants of the original tranplants matured at an earlier age, were larger, and also produced fewer and larger offspring. After 4 years males of the El Cedro stream had evolved the low predation strategy but the females had not. By 7.5 years the females matured later and at a larger size but had not produced fewer or larger young. Regardless, the changes were all in the predicted direction if natural selection was in operation, and it is very likely that in another year or two the females would have "caught up" with the males.

Reznick compared the rates of change in the guppy populations with changes that have been reported in laboratory experiments and in the fossil record. The standard unit in which evolutionary rates are measured is called the *darwin,* introduced by the British mathematician J. B. S. Haldane many years ago. It is equal to the natural log (*ln*) of a given measurement (*X*) taken at the end of an experiment minus the log of a measurement taken at the beginning of an experiment divided by the interval length (*dt*) in years:

$$\text{darwin} = \frac{\ln X_2 - \ln X_1}{dt}$$

The rates of change for measurable quantities found by Reznick ranged from 3700 to 45,000 darwins. These rates were about the same as those for other modern selection experiments and from four to seven orders of magnitude greater than rates reported in the fossil record. However, as paleontologist Philip Gingerich has shown, these rates can only be compared over similar interval lengths or must be rescaled to a common interval length. So the comparison with rates in the fossil record over millions of years is not too instructive, because the latter rates are averages of many selection events, some of which may counter each other over long periods of time. The beauty of Reznick's work is that many carefully constructed studies, all of a quantitative nature, clearly document how natural selection can work over brief periods of time to cause evolution in natural populations. These results can then be extrapolated to the fossil record. The rates of change revealed in the guppy experiments are sufficient to account, given the time involved, for all the macroevolutionary patterns in other organisms documented in the fossil record. This is not meant to imply that natural selection among individuals is, in fact, responsible for all macroevolutionary patterns (trends), only that it occurs rapidly enough to account for it. One of the great challenges for evolutionary biologists in the future is to figure out ways to test for the effects of higher level influences, such as species selection, on evolution.

Plastic Sticklebacks

The ancient Greeks believed that the form of organisms was fixed; an animal's place in the nature of things was unalterable. It is hardly surprising then, that despite Aristotle's modern somewhat logical approach to the world, neither he nor any of the other classic Greek philosophers ever developed a theory akin to evolution. It can be argued that in some ways the classical Greeks set back evolutionary biology more than 2000 years, until 1859, when Darwin published the classic *Origin of Species*. Yet Aristotle's thinking still permeates western thought, mostly through the Biblical fundamentalist interpretation of the "Ladder of Perfection," an essentially Greek construct representing the animal world with humans at the apex and less complex organisms than humans occupying lower rungs on the ladder. But the grand theory of evolution becomes more and more unassailable every day, and now we have evidence, not only for the process of natural selection through brief periods of time and a few generations but even within the lifetime of an individual.

Three-spine sticklebacks are small fishes that inhabit a number of lakes in the Old and New World. Recently, Dolph Schluter and other Canadian biologists have been studying an interesting series of stickleback species from small coastal lakes in British Columbia. The species have only recently been recognized, and they do not yet have formal species names. They are members of the genus *Gasterosteus*. Two of the species, when found together (e.g., in Lake Paxton), display significantly different morphologies. The biologists refer to them as the "benthic" and "limnetic" species, reflecting their preferred areas for foraging. Benthic organisms spend most of their time on the bottom, whereas limnetic species feed and live mostly near the surface. The benthic species has a relatively deep body, few gill rakers (tough extensions from the gills used for straining food from the water), and a wide mouth terminating in line with the central axis of the body ("terminal" mouth). The limnetic has a relatively narrow body, numerous long gill rakers, and a narrow upturned mouth. The limnetic species feeds more on plankton than the benthic species. These differences have been shown to be inherited. Both species are closely related and probably evolved from an ancestral marine species that colonized the lake during the late Pleistocene. Geological evidence suggests that these species could not have coexisted for more than 13,000 years. Solitary species found in other lakes (e.g., Cranby Lake) display morphologies intermediate to the benthic and limnetic types. Schluter and colleagues were interested in determining the answers to two ecological/evolutionary questions: Could the differences in morphological structure between the species be the result of character displacement of one species away from the other? How much of the morphological differences, especially in mouth structure, could be due to diet-induced phenotypic plasticity?

Before we go on, let us clearly define these principles. Character displacement is a hypothesis introduced by W. Brown and E. O. Wilson in 1956. They suggested that competition was an important force in structuring animal communities, and they further proposed that competition between species with considerable ecological overlap could lead to morphological differences between them in the geographical area where the two species overlap. Any area of overlap between two populations is called an area of *sympatry;* species there are *sympatric.* Species may also be *allopatric* for parts of their ranges; that is, there are areas where the two species are not found together. The hypothesis of character displacement predicts that two closely related species with similar ecological requirements will be more different from each other in the area of sympatry than they will be when allopatric (when geographically separate).

As we saw in Chapter 2, anatomical traits (phenotypes) are determined by the interplay of a number of influences, including heredity, development, and the environment. We have known for some years that certain organisms, mostly plants and invertebrates, could change forms depending on the habitats in which they were reared. But it has only been recently that we have become aware that some anatomical configurations in vertebrates, with a hard internal skeleton, could also be significantly modified in this manner.

In studies published in 1992, Schluter confirmed the *pattern* of character displacement, showing that the limnetic and benthic species were most different from each other when sympatric (**Fig. 12.6**). However, they had no idea how long it took to develop these differences nor what kinds of selective regimes were necessary to promote them. Consequently, they set up an experiment in artificial ponds at the University of British Columbia. The study was confined to a single generation of fish. They used a solitary species from Cranby Lake, with intermediate morphology, as the experimental "target."

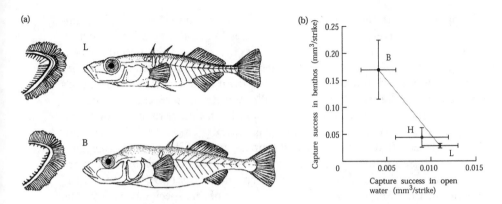

Figure 12.6 (a) The limnetic (L) and benthic (B) forms of the three-spined stickleback, showing differences in body form, mouth size, and gill raker length. In (b) we see that prey-capture success was greatest for each form in its preferred habitat (points lying closest to dashed line). Hybrids (H) fared worse than either parental form, thus reinforcing the hypothesis that natural selection is maintaining the separate morphs, which breed true. (From D.J. Futuyma, 1998. *Evolutionary Biology*, 3rd ed. Sunderland, MA: Sinauer Associates. p. 546; after D. Schluter, 1993. Adaptive radiation in sticklebacks: size, shape, and habitat use efficiency. *Ecology* 74:699–709.)

The limnetic species from Paxton Lake was used as the potential "competitor." Two ponds were divided in half and a young population of the target species (made up of the Cranby species and crosses between the Cranby species and the benthic species from Lake Paxton, to exaggerate anatomical differences away from the limnetic species) was evenly distributed in both sides of both ponds. Young of the Paxton Lake species were then introduced into one half of each pond (the experimental unit). The other side of each pond, with only the Cranby young, constituted the experimental control, against which results from the competition could be judged. The investigation found that those Cranby young most similar to the limnetic competitor suffered a significant depression in growth rate in both experimental populations. The controls demonstrated no such depression. In one of the experimental halves, survival of the Cranby phenotypes most similar to the introduced competitor significantly decreased, but it was not so reduced in the other experimental population. Although additional experiments are needed to confirm their findings, Schluter's work constitutes the first experimental evidence that competition can promote rapid directional selection, leading to character displacement (**Fig. 12.7**).

A second set of experiments was confined to the benthic and limnetic species from Lake Paxton. As noted above, the benthic species has a terminal mouth with a wide gape for feeding on larger prey items on the bottom, whereas the limnetic species has a narrow upturned mouth and feeds mostly on small planktonic organisms in the water column. Schluter's research team

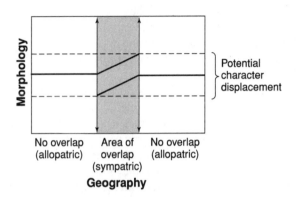

Figure 12.7 The concept of character displacement. Character displacement is identified when two species are different in one or more attributes (size and/or morphology) when together (sympatric) and more similar when apart (allopatric). One or both species may respond. Character displacement theoretically minimizes competition and is the result of natural selection against intermediate individuals.

reversed the diets of the species in populations reared in the laboratory and compared their morphology to control populations fed diets they would encounter in nature. They found that all but one of the traits they measured showed convergence toward the opposite morphotype; that is, young limnetics developed morphologies like benthics and vice versa. They determined that up to 50% of the overall morphological difference between the limnetic and benthic species can be determined by this "phenotypic plasticity" and concluded that such plasticity plays an important role in the overall ecological/evolutionary strategy of sticklebacks.

Lizards Without Limbs

One of the most successful radiations in animal history is the dispersal and subsequent adaption of *Anolis* lizards in the New World tropics. Including the dime-store "chameleon," *Anolis carolinensis,* these small lizards are regular inhabitants of the southeastern United States and throughout the Caribbean islands. Particularly on larger islands of the Greater Antilles, anole species are distributed according to perch type, body size, and climatic conditions. Anoles that perch on narrow branches are smaller and have shorter hindlimbs than those that perch on wider branches. There is a distinct trade-off here of speed and agility; the anoles on small perches are slower but more agile than those that inhabit larger trees. Recently, evolutionary biologist Jonathan Losos and his colleagues reported the results of a 10- to 14-year experiment with the anole *Anolis sagrei.* This species, which may have evolved in Cuba, occurs naturally in Jamaica and in the Bahamas. The investigators introduced *A. sagrei* to 14 very small islands in the Bahamas. All the lizards were taken from a natural population on a small island in the Bahamas named Staniel Cay. The experimental introductions began in 1977. The islands varied considerably in size and vegetational cover. Some were almost devoid of vegetation and others had a large vegetated area. The islands were stocked with either 5 or 10 individuals. Populations on the smallest islands disappeared quickly, within weeks. Eleven populations survived and expanded, and Losos and colleagues found that, amazingly, the different populations had morphologically diverged in a mere 10 to 14 years.

The observed changes mirrored exactly the expectations of morphology given the type of vegetation on each island, as compared with the founder stock on Staniel Cay. Most of the experimental islands have scrubby vegetation with narrow-diameter perches. Measurements taken on limb size showed that within this brief study period, the island populations had developed shorter hindlimbs and wider toepads, exactly the same modifications seen in other species of *Anolis* on islands with similar vegetation. This is an important study, because it shows how quickly a complex vertebrate animal can change under natural conditions. The rate of change ranged to more than 2000 darwins, which is extraordinarily fast, but actually below the range of rates reported for Reznick's guppies during approximately the same length of time (see above).

Although the changes in body form of the anoles are real and certainly demonstrate directional modification influenced by the environment, the investigators cannot determine yet if the changes were due to differential survival among individuals of slightly different morphologies (natural selection) or anatomical modification of individuals during their lifetime (called "ecophenotypic" change). We know that many organisms can actually change their anatomy somewhat depending on the environment in which they are found. For example, oysters grown on different types of hard or soft ocean bottoms will have different shell patterns. Certain plants will have different leaf shapes depending on whether they are grown in water or on land. Similarly, it is possible that the changes that occurred among the lizards were of this type; young lizards that grew up on stunted vegetation grew shorter limbs. This is in contrast to another model of change, where lizards of all morphologies are born, but those with shorter limbs reproduced more frequently than those with longer limbs, or those with longer limbs died for one reason or another while those with shorter limbs survived. This would be a natural selection scenario, where those best fit survived to reproduce. The jury is currently out on the two possibilities, but in any case one thing is clear: Regardless of the mechanism involved, evolution can progress at a very rapid rate.

Suggested Readings

Brown, W. L. and E. O. Wilson. 1956. Character displacement. *Systematic Zoology* 5:49–64.

Corbet, G. B. 1975. Examples of short- and long-term changes of dental pattern in Scottish voles (Rodentia; Microtinae). *Mammalian Review* 5:17–21.

Day, T., J. Pritchard, and D. Schluter. 1994. A comparison of two stickle-backs. *Evolution* 48:1723–1734.

Grant, B. R. 1985. Selection on bill characters in a population of Darwin's finches: *Geospiza conirostris* on Isla Genovesa, Galapagos. *Evolution* 39: 523–532.

Grant, B. R. 1993. Evolution of Darwin's finches caused by a rare climatic event. *Proceedings of the Royal Academy of London, Series B* 251:111–117.

Grant, B. R. and B. R. Grant. 1995. The founding of a new population of Darwin's finches. *Evolution* 49:229–240.

Johannesson, K., E. Rolan-Alvarez, and A. Ekendahl. 1995. Incipient repro-ductive isolation between two sympatric morphs of the intertidal snail *Littorina saxatilis. Evolution* 49:1180–1190.

Johnston, R. F. and R. K. Selander. 1971. Evolution in the house sparrow. II. Adaptive differentiation in North American populations. *Evolution* 25:1–28.

Losos, J. B., K. I. Warheit, and T. W. Schoener. 1997. Adaptive differentia-tion following experimental island colonization in *Anolis. Nature* 387:70–73.

Madsen, T. and R. Shine. 1992. A rapid, sexually selected shift in mean body size in a population of snakes. *Evolution* 46:1220–1224.

Rensberger, J. M. and A. D. Barnosky. 1993. Short-term fluctuations in small mammals of the late Pleistocene from eastern Washington; pp. 299–342. In R. A. Martin and A. D. Barnosky (eds.), *Morphological Change in Quaternary Mammals of North America.* Cambridge University Press, Cambridge.

Reznick, D. N., F. H. Shaw, F. H. Rodd, and R. G. Shaw. 1997. Evaluation of the rate of evolution in natural populations of guppies (*Poecilia retic-ulata*). *Science* 275:1934–1937.

Schluter, D. 1994. Experimental evidence that competition promotes diver-gence in adaptive radiation. *Science* 266:798–800.

Williamson, P. G. 1981. Paleontological documentation of speciation in Cainozoic molluscs from Turkana Basin. *Nature* 293:437–443.

Wood, C. C., and C. J. Foote. 1996. Evidence for sympatric genetic diver-gence of anadromous and nonanadromous morphs of sockeye salmon (*Oncorhynchus nerka*). *Evolution* 50:1265–1279.

Epilogue: Charles Darwin and the Fossil "Problem"

Darwin was well aware when he wrote *The Origin of Species* that there were some "difficulties" (as he put it) with the theory of evolution by natural selection. One of the positions that was of most concern to him was the almost complete absence of intermediate, or transitional, forms among the living world fauna and from the fossil record. He asks in Chapter 6, for example, ". . . why, if species have descended from other species by fine gradations, do we not everywhere see innumerable transitional forms?" And later, in Chapter 10, he wrote, "Why then is not every geological formation and every stratum full of . . . intermediate links? Geology assuredly does not reveal any such finely-graduated organic chain; and this, perhaps, is the most obvious and serious objection which can be urged against the theory."

Before I outline the problem and its solution in more modern form, let us once again be certain we understand what is meant by the word "transitional" or "missing link." Confusion about the definition of these terms can lead to significant misinterpretation. Darwin used the term transitional to mean any intermediate step in an evolutionary history in which differences between ancestors and descendants would be observed. The differences could be rather minute, like a slightly longer tail of a bird species as it evolved toward larger size, or considerable, such as the eventual loss of teeth in modern birds from their reptilian ancestors. When most people think of missing links, they are thinking of the latter kind of changes. So there is a problem of scale here that must be kept in mind. Although some animal characteristics may have evolved very rapidly (e.g., hairlessness in the naked mole rat, *Heterocephalus glaber,* very likely a single mutation that could have caused the absence of fur in a single generation), others may have taken thousands or millions of years to develop (e.g., the evolution of an ancestral bipedal dinosaur to a modern bird, through a number of intermediate steps).

Now back to the fossil "problem" as it existed in 1859, restated in slightly more modern form:

1. If evolution proceeds gradually, with successive generations differing slightly from preceding ones, all of nature should be in some sort of intermediate state, and we should see many transitional forms among living groups of organisms.

2. If evolution proceeds gradually, with successive generations differing slightly from preceding ones, we should see many fossil transitional forms in ancient sedimentary strata.

Many creationists, such as Philip Johnson, author of *Darwin on Trial*, use Darwin's concerns and the 1972 hypothesis of punctuated equilibrium developed by Niles Eldredge and Stephen J. Gould (see Chapter 2) as part of their arsenal to prove that evolution is false. Eldredge and Gould suggested that one of the reasons Darwin and others failed to see lots of gradations of intermediate forms in the fossil record is that evolution does not proceed gradually. Rather, most species change little during their geological life spans, with their morphology just meandering around an average condition, sometimes for millions of years. The changes that do occur happen very quickly in a geological sense, probably less than 1% of a species' life span and only during the initial time of branching of a descendant from its ancestor, the "speciation event." Creationists suggest that the punctuated equilibrium model of evolution is simply an attempt by evolutionists to argue their way out of the absence of transitional forms while preserving the theory of evolution. So, how do we answer both Darwin's concerns and the creationist position, incorporating the punctuated equilibrium perspective? Well, to begin with, Darwin was no dummy. He answered most of the concerns himself, almost 150 years ago, in the same book where he first gave them voice. In short, the responses take the following form:

1. If natural selection operates to improve species, that is, to lead them toward better adaptations to their regional environment, then it stands to reason that those generations and species that are better suited for their environments will replace earlier organisms that were not as well adapted. The mechanism is competition, which results ultimately in reproductive success of certain individuals and species and the extinction of those less fit.

2. The absence of perfectly intermediate species among living groups of species is in part because none of the group's members may have evolved from each other. Rather, each will show similarities through ancestral generations back to one or more common, now extinct, ancestors. Intermediates are in the fossil record, not among living animals.

3. The absence of continuity among living species is at least in part related to the fact that the physiography of the Earth is constantly changing, and therefore it is clear there have been physical gaps separating species; indeed, these gaps may have led to their formation in the first place.

4. The absence of many examples of transitional forms in the fossil record has been, and continues to be, mostly a function of the incomplete nature of the fossil record and the very low probability that transitional forms will be discovered. This is primarily due to the following:

    a. Absence of appropriate sediments. Most of the Earth's ancient crust is gone, having been weathered away by erosion. Sedimentary basins are often very localized, and even if a sequence of fossiliferous localities in these basins spans a considerable period of time, it is in a very restricted geological area and the probability that speciation events or transitions of any kind would occur exactly there is infinitesimally small.

    b. Much of the Earth's sediments are covered by vegetation or water.

    c. Minimal time spent searching the rocks for fossils by a limited number of people; paleontology is not a field that receives much funding as compared with, for example, cancer research.

5. The absence of transitional forms in the fossil record has partly been the result of an old typological species concept in which paleontologists were more concerned with naming new species than with determining their relationships. Consequently, it has only been within the last 50 years or so that the continuity between some fossil forms has been recognized.

6. Studies of fossil organisms since Darwin's day have shown a tendency, in many groups, for species to remain morphologically stable over long periods of time; the "equilibrium" in the punctuated equilibrium hypothesis of Eldredge and Gould. This would, as Eldredge and Gould maintained, reduce tremendously the probability of finding transitional forms between species because evolutionary change would be restricted to a very brief period at the beginning of a species' life span.

There is an element of truth in Philip Johnson's rhetoric, namely that most of these points add up collectively to excuses why transitional forms are rare in the fossil record. Let us consider the worst case scenario. Suppose that all six points are indeed true; that evolutionary change occurs very rapidly, in short bursts, and therefore the probability really is very small that we will uncover transitional populations between species, and fossiliferous sediments are as rare as hen's teeth. Can the fossil record then reveal that evolution has occurred? Fortunately, we do not need transitions between species to prove that evolution has occurred. The use of the punctuated equilibrium model in creationist argument is a straw man, for a couple of reasons. First, there is this matter of scale; the really large transitions appear to take hundreds of thousands to millions of years to play out, not the paltry few thousand years of a speciation event. Transitions between species are not expected to demonstrate much significant anatomical change. All we need from the fossil record is part of the full evolutionary sequence of a large scale transition to prove that it occurred. That is, there must be a documented series of terrestrial animals that precede semiaquatic ones that eventually lead to aquatic whales, or small horses with three front toes occurring in sediments older than larger ones with one toe, and so on. The probability of finding fossil hominid skeletons is extraordinarily remote, yet 143 years after Darwin's *Origin of Species* we have a bunch of them, and there are no contradictions. Ape-like forms precede upright forms with small brains, which precede upright larger forms with large brains. And the fossil record is consistent with modern information on the genetics of humans, apes, and monkeys.

The Baron Georges Cuvier, a great late eighteenth and early nineteenth century anatomist and arguably the father of paleontology, was the first to seriously inquire as to the nature of fossil animals by comparing them with modern animals. This approach, known as the comparative method, is the basis for all of our knowledge of the relationships of fossil organisms. Comparisons are made between ancient and modern animals and the extent of their similarities and differences determines the degree of relationship and thereby how they are classified in the great tree of life. Although he practiced this logical approach with fossils in the Paris Basin of France, Cuvier nevertheless concluded that entire sets of marine species had been wiped out and then replaced by the Deity. His model for the sudden replacement of fossil communities is today called "catastrophism." Even the more erudite of creationist writers, such as Philip Johnson, like to conflate Eldredge's and Gould's punctuated equilibrium hypothesis with catastrophism. It makes some sense; if transitional forms do not exist (though Eldredge and Gould never said that) and punctuated equilibrium is just an excuse for their absence, then maybe the brilliant Cuvier was right: God creates and God destroys. Despite his prodigious intellect, Cuvier had a huge blind spot; he did not know how to recognize the results of erosion in the geological record, and as a result he misunderstood the meaning of superposed assemblages of different fossil marine organisms. Modern geologists (and most school children) now understand that the different assemblages observed by Cuvier are separated by missing intervals of time, when sea level was lower and erosion destroyed the intervening rocks and organic remains. When the sea level rose again, new groups of organisms colonized the same area. It is not difficult to understand how even the brightest of eighteenth century scholars, steeped in the religion of the time and working with a very limited knowledge of geology, could make such an error.

So we have come full circle. In the Introduction I tried to demonstrate that all living organisms are in one sense missing links, because they can trace their family history back to a common ancestor some 4.5 billion years ago. And the process that leads from a dinosaur to a bird is no different from one that leads to the origin of any species; it just plays out over a longer period of time. To truly understand the metaphorical miracle of evolution, one must observe the resulting patterns at many scales, much as in one of Benoit Mandlebrot's fractals. Individuals and entire species compete for resources (mostly energy in the form of food), and eventually one species replaces another. As the environment changes (and perhaps even without environmental change), species change, trying to keep pace. Careful study of dense fossil records in single sedimentary basins shows that species in most communities turn over constantly; they are not replaced as a unit. If Van Valen's Law of Constant Extinction dominates in the history of life, extinction is inevitable regardless of evolved adaptations, and it seems to play out faster in smaller than in larger animals (as does origination). Species originate as small peripheral isolates from other species, though in some cases (sympatric speciation) they may originate within an ancestor's species range if they are small and very restricted in their habitat choices. The slight differences between ancestor and descendant species may become exaggerated in thousands through millions of years, seen as phyletic change, or phyletic evolution within lineages. Through the genetic processes of mutation and recombination, new features continue to appear at random. Some may be adaptive, some are not. Given the immense time since the origin of life, it is highly likely that some evolutionary changes occurred very rapidly, especially where the characters are discrete versus continuous (e.g., the number of nipples in female mammals or hairlessness in naked mole rats). Every population of every species is an experiment in evolution. When we think of the millions of experiments going on daily in a colony of viruses or bacteria, it is no wonder that we continue to select for microbes that are resistant to our antibiotics. Over vast periods of time the full story of evolution is told only by paleontologists, and the patterns they perceive, of species extinctions and originations and of anatomical change, can only be the larger scale patterns of the evolutionary fractal. The vast majority of organisms that have lived are unavailable for study. They die and their remains are decomposed in ecosystems and used again in elemental form. Only under very special circumstances are organisms preserved as fossils. Even then these burial grounds are mostly destroyed by the erosive action of wind and rain, freezing and thawing, the meandering of rivers, the rising and falling of sea level, and other natural forces. We are left with precious little of our planet's history, but it is a testament to the universality of the evolutionary process and the tireless dedication of Earth's paleontological community that the evidence is there and that it is unmistakable. It may come as a shock, but Charles Darwin was no atheist, and he summed it up best in the last sentence to *The Origin of Species:*

> *There is grandeur in this view of life, with its several powers, having been originally breathed by the Creator into a few forms or into one; and that, whilst this planet has gone cycling on according to the fixed law of gravity, from so simple a beginning endless forms most beautiful and most wonderful have been, and are being evolved.*

G

GLOSSARY

abiotic: nonliving part of the environment, such as water, heat, rocks, etc.

absolute dating: dating of rocks or organic material contained in them that results in real-time ages for the units; such as radiocarbon dating.

adders: a group of poisonous snake species in Eurasia and Africa.

allele: form of a gene.

allometry: the disproportionate growth of structures in the body or within the same organ that can result in significant physical change or evolutionary change.

allopatric: separated geographically.

alveoli: sockets in palate and lower jaw in which roots of teeth fit (*alveolus* is singular).

amniote: vertebrate animals (reptiles, birds, mammals) that have an egg or placental connection including certain membranes, such as the amnion and chorion.

amphicoelous: condition of vertebrae in which both ends are concave.

amplexus: mating behavior of modern amphibians, in which the male grasps the body of the female, resulting in the female laying eggs and the male releasing sperm.

anadromous: organisms that live most of their adult lives in the oceans but migrate into brackish water or fresh water to breed.

anagenesis: the origin of a new species by phyletic evolution. The beginning and the end products of the lineage are considered to be different enough to be named as different species. Not accepted by those scientists, such as the author of this book, who consider the origin of a new species to be the result only of branching, or cladogenesis.

analogous: structures in different species that share the same function but not the same ancestry, such as the wings of bats and butterflies.

anerobic: lack of oxygen; or organisms that do not use oxygen for respiration. Characteristic of some bacteria in hot, sulphurous springs and deep ocean vents.

apomorphy: a derived (advanced) characteristic.

arboreal: pertains to biological activity in trees.

autapomorphy: a characteristic used to diagnose a taxon as unique from every other taxon of like rank.

Beagle, H. M. S.: the ship on which Charles Darwin sailed around the world.

belief: a hypothesis that has not been, or cannot be, rigorously tested.

bell curve: graphic representation of a Normal distribution, in which measurements become symmetrically less frequent the farther they are from the average, or mean value.

Bering land bridge: a land connection between Siberia and Alaska that occurred at various times during the past when sea levels were considerably lower.

Big Bang: hypothesis that all matter in the universe was created in a huge explosion around 12 to15 billion years ago.

binomial: genus and species names, as in *Homo sapiens.*

biogeography: the study of the distribution of organisms and the processes that led to those distributions.

biological species concept: the concept, proposed by Ernst Mayr, that species are defined by their reproductive (gene flow) limits.

biosphere: the atmosphere, surface, and subsurface area of the Earth that supports life.

biostratigraphy: the determination of the relative position of rock layers by the fossils contained therein.

bipedal: walks on two feet.

bottleneck: a form of genetic drift caused by the near extinction of a population, drastically reducing its genetic variation.

brachiation: form of arboreal locomotion: hand-over-hand swinging from branch to branch.

brachydont: low-crowned, referring to teeth.

bunodont: flat-crowned, referring to teeth.

cement(um): hard substance that evolved in the teeth of many animal groups through time. Provides rigidity during chewing (mastication).

cenogram: a graphic plot of body mass on the *y* axis against rank order of mass on the *x* axis. Used to infer the nature of past habitats.

cheek teeth: teeth in reptiles and mammals posterior to the canine.

chemosynthetic: the use of energy in chemical elements instead of plants (photosynthetic) as the energetic basis for an ecosystem. Seen today among organisms around deep-sea hydrothermal vents.

choanae: internal nostrils of vertebrates.

chron: major period of time when the Earth's magnetic field is either as it is today (normal) or was reversed. Shorter periods are termed *subchrons.*

chronomorph: informally named populations of a species, usually restricted to a range of time in a species' life span, demonstrating one or more unique structural features or size differences from other populations in the same species.

clade: an ancestor and all of its descendants; a monophyletic group.

cladistics: the scientific study of the phylogeny of organisms, resulting in a graphic representation known as a *cladogram.*

cladogram: graphic representation of the relationships of organisms determined by phylogenetic analysis.

clavicle: collarbone of humans. Right and left clavicles are fused into the furcula, or "wishbone" of birds.

cochlea: spiral structure of the inner ear that conducts electronic impulses to the auditory nerve to be interpreted as sound.

coelacanth: common name for species in the genus *Latimeria,* a living actinistian lobe-fin.

coevolution: theoretical concept suggesting that some species pairs or groups, such as parasites and their hosts, evolved through long periods of time with each other.

community: assemblage of species in an ecosystem, such as the "snake community" or the "shrub community." Can include any groupings of organisms.

composite: as used in paleontology, referring to the condition in which a fossil, or a slab on which a fossil is found, is made up of two or more pieces cemented together.

continental drift: a theory that the world's continents have moved in the past. Supported by the science of plate tectonics.

convergent evolution: similarities in structure evolve among distantly related organisms to solve the same ecological problems. Example: fusiform body shape in porpoises and sharks for rapid, predatory movement in water.

Copes Law: the observation that in most clades there is more of a tendency to evolve toward larger size than the opposite.

cosmology: the study of the origin and nature of the cosmos, the rest of the universe far from the Earth.

creationism: the idea that the universe and the world were created by an omnipotent supernatural being called God.

crown: part of vertebrate tooth composed of an outer enamel layer surrounding a dentine center.

cusp: protuberance from the crown of the teeth in vertebrate animals. Shape is often highly correlated with diet.

deep time: the concept that the Earth and the universe are billions of years old.

deme: a breeding group within a population.

diapsid: condition of vertebrate skull in which two large holes, or fenestrae, are located at the back. Early vertebrates, particularly some of the reptiles, were in the past partly classified on the basis of these holes and where they were located. Turtles, with the anapsid condition, lack these holes. Most early theropods are diapsid, as are the earliest birds.

diastema: space between teeth, can be in upper or lower jaw. Pronounced in rodents.

differentiation: refers to the condition of molar enamel thickness in arvicolid rodents (see Chapter 7).

diploid: double, or adult, number of chromosomes in most organisms.

directional selection: natural selection that acts on individuals to shift the average value of a measured variable (characteristic) either up or down.

disruptive selection: natural selection that acts against individuals with the average value of a measured variable, resulting in a distribution with two new means; a bimodal distribution.

diversity: measure of the number of organisms in an ecosystem. Sometimes represented by a combination of numbers of species and the distribution of individuals in those species, but more often simply as the total number of species (see *species richness*).

Dollo's Law: a theory stating that evolution never reverses.

Doppler effect: displacement of waves (sound, light) indicating direction of movement of an object relative to an observer. Red-shifted objects are moving away, blue-shifted objects toward, an observer.

Down syndrome: a genetic abnormality in humans, caused by nondisjunction of chromosome number 23, resulting in a variety of defects, including mongoloidism.

Drosophila: a genus of fruit flies including many species. Used in genetics experiments because of their short generation time (a few weeks).

ecophenotype: form of a species that is variable depending on the environment in which it is found. The difference is not due to different genes.

ectothermy: a condition in which a body temperature of an organism follows that of the environment.

endothermy: the condition in which an organism is able to create a body temperature above that of ambient (environmental) by physiological means alone. The only living endotherms are birds and mammals.

epicontinental: used in the construct "epicontinental sea," referring to shallow seas that covered parts of what is today dry land on some of the major continents.

eukaryotes: organisms with a membrane surrounding their chromosome (nuclear membrane). Includes most everything except viruses and bacteria.

eurytopic: living species that can be found within a number of ecosystems.

experiment: a test in which a variable of interest that has been modified (experimental group) is compared to one in which it has not (control group).

fitness: a quantity that expresses the probability of survival of individuals in a population. Usually measured as some form of reproductive success.

fossil: usually applied to dead organisms in which some replacement by minerals has taken place in the remains.

founder effect: process by which a small number of individuals colonizes an area and establishes a new set of gene frequencies. One process of genetic drift.

fundamentalism: a religious perspective that includes a belief that the religious books, such as the Christian Bible, are the word of God and are, therefore, true in every detail.

genetic drift: random distribution of individuals with partial sets of genes that can result in new populations that have different appearance (phenotype) from the parent population.

genotype: the genetic combination that underlies a certain physical feature (phenotype).

glaciation: at various times in the past the Earth has cooled to the point that much of the Earth's water remained trapped for long periods as ice, both at the poles and on mountain tops. As the ice accumulated, massive continental glaciers moved southward in the northern hemisphere and northward in the southern hemisphere scouring valleys. These advances were accompanied by significant drops in sea level. Ice masses known as alpine glaciers also moved down from mountain tops.

greenhouse gases: gases in the atmosphere, such as carbon dioxide, that allow solar energy to pass through and help hold in heat.

half-life: the amount of time it takes for half of an isotope to convert to a daughter product.

haploid: half of the adult chromosome number. Found in sex cells, the gametes.

heterochrony: any change in rate of development.

home range: the geographic area in which an animal satisfies its biological needs. Does not include migratory range. Known to be positively correlated with body size.

homeotic genes: master genes that control a variety of important developmental processes in a wide variety of organismal groups.

homologous: similar underlying structures in different species are explained as the result of sharing a common structural ancestry, such as the feet of horses and the feet of earlier mammals that have not been modified to the same extent.

homoplasy: circumstance, from cladistics, in which two features are similar through evolutionary convergence rather than close phylogenetic relationship (see *convergent evolution*).

hopeful monster scenario: idea proposed by geneticist Richard Goldschmidt suggesting that many new species have arisen through mutations that changed their features relatively quickly, perhaps in a single generation. Not widely accepted at the present time.

Hubble constant: a value that indicates the rate at which extraterrestrial objects, such as stars, are receding from Earth and from one another. Around 15 km/sec per million light years (mly). That is, an object 1.0 million light years away from Earth is moving at a speed of 15 km/sec; at 2.0 mly it would be travelling at 30 km/sec.

hybrid zone: a limited area in which two populations, normally considered to represent different species, can interbreed.

hypothesis: an idea, in science usually based on observations.

hypsodont: high-crowned, referring to teeth.

index species: a fossil species that is characteristic of a particular time range in the past.

indicator species: a living species found in a fossil assemblage that, because of its limited habitat choice (stenotopy) and physiological tolerances, provides proxy information on past climate in the area of deposition. If the species is found as a fossil outside its known historical range, it is called a *disjunct* species.

infrared: part of electromagnetic spectrum that produces heat.

insolation: the radiation environment at the Earth's surface; usually measured as a rate of energy transfer in $cal/m^2/year$ from the sun.

interspecific: between different species, as in interspecific competition.

intraspecific: within the same species, as in intraspecific competition.

iridium: rare metallic element, the primary source of which is extraterrestrial. A heavy concentration of iridium in ancient sediments can indicate the presence of an asteroid impact.

isomer: mirror image geometric forms of chemical compunds, such as the L and D isomers of amino acids.

isotopes: stable or unstable variants of elements, often used in radiometric dating or other studies, such as stable oxygen isotope work. Examples include ^{14}C or ^{18}O.

Ladder of Perfection: also known as "ladder of progress," "chain of being," etc. The concept, which can be traced to the classic Greeks, says that species are immutable and that some species, such as humans, are "more perfect" than others and thereby sit higher on the ladder. A religious concept followed today by fundamentalists.

lagerstatten: "mother lode" in German. Implies a very special, rich fossil deposit.

lateral line: system of receptors in aquatic tetrapods, usually running from the head along the midline of both sides of the body, that picks up vibrations.

Law of Constant Extinction: described by Leigh Van Valen in 1973, this principle suggests that all groups of organisms go extinct at rates characteristic for each group (see *Red Queen hypothesis*).

lichen: a symbiotic arrangement between a fungus and an alga.

light-year: the distance light travels in one year at the speed of light.

lineage: the history of a single species.

liposomes: spherical structures that spontaneously form from lipid molecules in simple lab experiments. May indicate how earliest cell membranes evolved.

lithosphere: surface of the Earth.

locus: location of a gene on a chromosome.

macroevolution: large scale patterns of evolutionary change revealed over long periods of time, usually hundreds of thousands or millions of years.

maternal effect: association with the mother can determine the phenotype (appearance) of the offspring to some extent.

meiosis: reduction division. A process occurring in the gonads resulting in the haploid number of chromosomes in the sperm and the egg.

melanic: dark, as in a dark form of a butterfly.

metabolic rate: usually measured as oxygen consumption, it represents the rate at which an organism processes energy. In endotherms (birds and mammals) it is also the source of body heat.

metazoans: advanced life forms with distinct organ systems.

microevolution: evolution within species, or lineages over relatively brief to long periods of time. In some cases, considerable change can accumulate in lineages by this process.

monodactyl: runs on one digit, as in modern horses, as opposed to tridactyl, the condition seen in some ancient horses.

monophyletic group: a single ancestral species and all of its descendants.

morphospecies: species, such as those in the fossil record, diagnosed exclusively on physical characteristics, for which no breeding information is available.

morphotype (or **morph**): one of a number of anatomical forms.

mosaic evolution: unequal evolutionary rates in the same or different characteristics in the same or different populations within one species. Results in different contemporaneous populations that may express either or both primitive and relatively advanced traits.

neutral characters: features of an organism that may not be under the auspices of natural selection. Debatable.

notochord: cartilaginous rod that provides rigidity and bilateral symmetry in chordate animals. Becomes reduced in size as skeletons appear, and is transformed during development into the centrum of the vertebrae in mammals. Embryonic notochord cells induce formation of the spinal cord during development.

null hypothesis: the standard "no difference" hypothesis of statistics, in which a sample or variable of interest (experimental group) is initially considered to be no different from the control group, and only if that hypothesis is rejected can they be said to be significantly different.

ontogeny: development from embryo until first reproduction.

opercular: gill region of fish-like vertebrate animals.

orthogenesis: ancient idea that organisms evolve in a linear fashion toward a goal of perfection (interpreted in either a philosophical or religious manner).

orthograde: vertebrate posture in which the spine is held vertical to the ground, as in brachiation by apes or human walking.

oviparous: egg-laying.

ovoviviparous: young hatch from eggs, but the eggs are retained in the body of the female.

paedomorphosis: a hypothetical evolutionary process that results in embryonic features being retained into adulthood.

paleoanthropologist: someone who studies the ancient history of humans.

paleomagnetism: the study of changes in direction of the Earth's magnetic field.

paleontologist: someone who studies ancient life, generally not including humans.

Pangaea: hypothetical supercontinent that once included all the continents of the world.

parallel evolution: closely related organisms evolve in different world areas, thus developing somewhat different features. Example: coniferous forests appear in both the Alps and the Rocky Mountains, but the species are often different.

parameter: in ecology, a variable (feature, characteristic). In statistics, it is a population feature, such as a population mean or standard deviation.

parsimony: concept implying that the simplest explanation or solution is likely to be the correct one. An underlying principle on which all cladistic analytical software operates.

passive competition: replacement of one species by another by producing more offspring.

peramorphosis: a hypothetical evolutionary process that results in new features being added during the later stages of development.

phenetics: a method of classification in which organisms are grouped strictly according to statistical distances between sets of characteristics. Often results in a classification that does not reflect phylogeny.

phenotype: the physical appearance of something.

phyletic evolution (change): change within lineages (single species).

phylogenetic analysis: an examination of the characteristics of organisms to determine their phylogeny, or genealogy. Usually incorporates particular computer software such as PAUP or MacClade to facilitate the process. Also known as cladistics.

phylogeny: the genealogical history of life or any part of it, often portrayed in a time framework.

phylotopic stage: time during development when the embryos of related organisms all look very similar because the basic patterns of development to this point are directed by the same homeotic genes.

phytoplankton: small, often microscopic, organisms found in aquatic ecosystems that can photosynthesize simple sugars.

ping-pong ball hypothesis: idea that simple life might have arisen in the solar system on one or more planets and been accidentally spread around, probably riding on pieces of rock blasted off planets as a result of asteroid or cometary impacts.

pinna: earlobe.

placental: mammals with a true placental connection between fetus and parent, as opposed to marsupial mammals, in which the embryo transfers itself to a pouch, or marsupium. There are also oviparous, or egg-laying mammals, such as the duck-billed platypus.

plasma membrane: cell membrane.

plate tectonics: the scientific study of the dynamics of the pieces that make up the Earth's crust.

plesiomorphy: a primitive characteristic.

polymorphic: variable.

prognathism: condition in apes and early hominids in which the face is extended forward into a muzzle-like configuration.

prokaryotes: organisms, such as bacteria, that do not have a membrane surrounding their chromosomes.

pronograde: vertebrate posture in which the spine is horizontal to the ground.

pseudoextinction: the disappearance of a named species in a phyletic sequence (lineage), within which two or more species have arbitrarily been named. Corresponds to speciation via anagenesis, which is not accepted as speciation by some scientists.

punctuated equilibrium: a hypothesis proposed by Niles Eldredge and Stephen Jay Gould that most species remain unchanged for most of their geological life span (the equilibrium). Change in species is restricted to a brief period during the speciation event that created the species in the first place (the punctuation).

pygostyle: last caudal, or tail, vertebra in birds. Supports the tail feathers and the preen gland.

random walk: a statistical concept suggesting that directional change in organismal features through time may be due purely to chance.

Red Queen hypothesis: the concept, proposed by Leigh Van Valen in 1973, that species try to keep up with the introduction of new competitor species and a changing environment but never quite succeed (see **Law of Constant Extinction**).

scale: different levels of inquiry, such as different time periods. For example, studies can be over brief periods of time (one scale) or over millions of years (another scale). The patterns observed may differ simply by virtue of the scale examined.

scientific method: an approach to knowing anything that includes the elements of observation, hypothesis, and testing (experimentation).

scientist: someone who uses the scientific method to solve problems.

Second Law of Thermodynamics: one of the laws of physics that, in biological form, implies that no energy conversion is ever 100% efficient. Some energy is always lost as heat.

secondary palate: bony roof of mouth in mammals; separates breathing and chewing functions.

sectorial: dentition, usually implying a carnivorous diet, in which the cusps of the teeth are sharp and pointed.

sexual selection: concept in which females are eventually responsible for the evolution of display structures in males (e.g., peacock feathers, gaudy colors) by preferentially choosing males with more elaborate structures for mating.

Signor-Lipps effect: the appearance of gradual extinction in the fossil record as an artifact of limited collecting. First described by paleontologists Philip Signor and Jere Lipps.

speciation, or speciation event (cladogenesis): the origin of a new species from a preexisting species. The net number of species increases. Also known as "branching."

species: individuals, often arranged in populations, that interbreed with each other but not with other related groups of individuals.

species richness: an ecological term; simply the total number of species in an ecosystem. Often used as a measure of diversity.

species selection: natural selection among entire species, rather than among individuals within species. Usually implies that there are "species level" characters, such as population size, that can affect competitive outcomes between species but are not found in the individuals themselves. Highly theoretical.

stabilizing selection: natural selection that acts to maintain individuals with the current average value of a measured variable (characteristic).

stable oxygen isotopes: isotopes of oxygen that do not decay into daughter elements.

stapes: or stirrup; one of the three middle-ear bones of mammals. First seen in Devonian tetrapods.

stenotopic: living species restricted in their distribution to a limited number of ecosystems.

sternum: breastbone. In birds, the presence of a keel indicates dynamic flight.

stratigraphy: the study of sedimentary rocks and their position relative to one another.

stratum: a layer of rocks. *Strata* is plural.

stromatolites: pillar-like aggregations of colonial cyanobacteria.

subduction: the movement of one continental plate under another. The major earthquake centers of the world occur along subduction zones.

superfecundity: an overproduction of offspring that leads to selection of those best fit to survive, or natural selection.

sympatric: in the same geographic area.

symplesiomorphy: a shared primitive characteristic.

synapomorphy: a derived (advanced) feature shared by two or more taxonomic groups.

systematics: the scientific study of the classification of organisms. Often incorporates phylogenetic analysis.

taxon: any formally recognized group of organisms; a species, a family, etc.

thanatocoenose: a death assemblage; an aggregation of long-dead organisms that does not necessarily represent the distribution or diversity of those organisms when they were alive.

thermoregulation: the manner in which an organism responds to its thermal environment.

tree of life: life's genealogy, or phylogeny.

trend: a directional change in a species or series of species through time.

tympanum: tympanic membrane of outer ear.

ungulates: hoofed mammals.

unguligrade: condition, seen in horses, in which locomotion is on the tips of the digits.

uniformitarianism: principle indicating that the features of the Earth in the past were shaped by the same ones that are at work today.

zygomatic arch: cheekbones in terrestrial tetrapods.

INDEX

Note: A page number in boldface indicates an illustration.

THE JONES AND BARTLETT SERIES IN BIOLOGY